W9-BCT-523

The Wind Masters

The Wind *Masters*

The Lives of

North American

Birds of Prey

PETE DUNNE

Illustrations by David Allen Sibley

A MARINER BOOK

Houghton Mifflin Company

BOSTON   NEW YORK

FIRST MARINER BOOKS EDITION 2003

For information about permission to reproduce selections from this book, write to Permissions, Houghton Mifflin Company, 215 Park Avenue South, New York, New York 10003.

Visit our Web site: www.houghtonmifflinbooks.com.

Library of Congress Cataloging-in-Publication Data

Dunne, Pete, date.
    The wind masters : the lives of North American birds of
prey / Pete Dunne ; illustrated by David Sibley.
        p.   cm.
    Includes bibliographical references.
    ISBN 0-395-65235-9
    ISBN 0-618-34072-6 (pbk.)
    1. Birds of prey—North America.   I. Title.
QL696.F3D868   1995
598.9'1'097—dc20   95-18314   CIP

Book design by Anne Chalmers
Text type: Sabon and Cochin (Adobe)

Printed in the United States of America

QUM 10 9 8 7 6 5 4 3 2 1

TO

ICARUS,

WHO SOARED,

AND TO CLAY SUTTON

(WHO HAS BEEN AROUND THE

BLOCK A FEW TIMES HIMSELF)

Contents

Contents

Acknowledgments

IT TOOK THREE YEARS to move this book from concept to completion. During this time many, *many* wonderful people lent their time, talent, and support. These include: Dennis and Wendy Allen, Sharon Bartles, Gus Daniels, Bob Dittrick, Dan and Melinda Droge, John Economidy, Ann and Bob Ellis, Laurie Goodrich, Kenn Kaufman, Clay and Pat Sutton, Sherri Williams, and Tom Wood, who provided logistical support.

Biologist Brent Bibles; Keith Bildstein, Ph.D., of Hawk Mountain Sanctuary; Prof. Eric G. Bolen, Ph.D., of the University of North Carolina; Paul Kerlinger, Ph.D.; Paul Lehman and Shawneen Finnigan of the American Birding Association; biologist Carol McIntyre; Prof. Mark Pokras, D.V.M., of Tufts University School of Veterinary Medicine; biologist Jay H. Schnell; Lisa Smith of Tri-State Bird Rescue; biologist Ted Swem, Ph.D., of the U.S. Fish and Wildlife Service, Fairbanks, Alaska; and Michael Wallace, Ph.D., of the Los Angeles Zoo, who offered technical assistance in the preparation of the manuscript or who reviewed chapters relating to birds falling within their fields of expertise.

Ned Harris, Tony Leukering, Ray Schwartz, Mitch Smith, and Clay Sutton, whose raptor photos were referenced for assorted illustrations.

Special thanks are reserved for Judy Toups, who cast her writer's eye over all chapters; Dr. Helen Snyder, who applied her skills as a raptor biologist to sift and leaven many chapters in this book; and my wife, Linda, and editor Dorothy Clair, who found time in their busy lives to review the work in its entirety.

The fieldwork of those biologists whose efforts laid the foundation of our knowledge concerning birds of prey have my special thanks and high regard. Grateful and deferential acknowledgment is *particularly* extended to the authors of several masterworks cited in the bibliography: Dr. Tom Cade, author of *The Falcons of the World*; Dr. Paul A. Johnsgard, *Hawks, Eagles and Falcons of North America*; Dr. Ralph S. Palmer, *Handbook of North American Birds*; and Drs. Noel and Helen Snyder, *Birds of Prey — Natural History and Conservation of North American Raptors*. Their books, which house a synthesis of biological data on North American raptors, made my task as storyteller an easy one.

Recognition would not be complete without acknowledging the debt I owe Tom Gilmore and the staff of the New Jersey Audubon Society, who gave a fellow staff member the latitude to shirk duties that they, in turn, assumed. In this regard, the staff of the Cape May Bird Observatory — Kathy Iozzo, Pat Sutton, Joan Walsh, and Louise Zemaitis — earn special consideration and my thanks.

This acknowledgment would not be complete without introducing and thanking editor Nancy Stabile, whose task it was to take a raw manuscript and convert it into a more readable form. She is a master of the editor's art, and for this readers can be grateful.

Acknowledgments

Introduction

IN YOUR HANDS ARE STORIES: fictional portrayals of nature's most celebrated aerial predators — the birds of prey. Like most stories, these are founded in truth (what a scientist might call facts), and from them readers will learn much about North America's thirty nesting species of diurnal raptors and the three vulture species.

But these accountings, one to a species, have a life beyond the disciplined standards that distinguish scientific treatments. That is what makes this book different from most other books about birds of prey. That is why I felt compelled to write it as I have.

In stories, Condors can dream, Red-tailed hawks can seek affection, and Goshawks can kill with pleasure. In this book you will meet a Black Vulture on his way to a picnic, a sexually frustrated Harris' Hawk, and a Raven who quotes poetry at the death of an eagle. In stories, as in love, as in war, all is fair.

Stories give a writer the latitude to explore possibilities that may go beyond the facts. If truth is a pond and facts stepping-stones, stories are the medium that allows writers to walk on water where the stones end.

Why write about birds of prey in this way? Why not just express the facts and facets of a raptor's life in an expository manner? After all, the birds we call the diurnal raptors are fascinating without literary elaboration. They have captured human imaginations, infiltrated our religions, become emblematic symbols of nation-states. They push our buttons in ways that few other groups of birds do.

The reason I have chosen the form of fiction to impart an understanding of an appreciation for birds of prey is simple: for effect, or, more precisely, for the effectiveness of this writing technique as an instructional tool. Hard information, I have found, is often more easily assimilated when it is presented obliquely. This is what makes parables and nursery rhymes and folklore so effective. This, and the pleasure of the telling, is why storytellers recite stories.

Twenty years ago, when I first grew fascinated by raptors, I purchased Arthur Cleveland Bent's two-volume treatise *Life Histories of North American Birds of Prey*. It was, and remains, an impressive accounting of lore compiled by a dedicated ornithologist, and the wealth of information contained defies the pages to contain it.

I read both volumes in their entirety. Read them in a four-day marathon, with pencil in hand and notes overflowing the margins. Read those 518 pages to a white-knuckle, index-finger-licking standstill.

Know what? When I finished the last chapter and considered the wisdom I had gleaned, I found that the profusion of facts had fused into an irretrievable mass. The only things I could recall were the accountings of contributing ornithologists whose personal experiences were used to impart information or to support facts. What stuck was the stuff imparted anecdotally, not didactically.

The Wind Masters tries to expand upon the anecdotal wisdom inherent in Bent. It tells stories about birds of prey and lets the facts that distinguish the various birds go along for the ride.

Because carriage is accorded as much importance as what is being carried, not all treatments of all species are substantively exhaustive. Sometimes, for example, a nest description may be missing in a particular species treatment; or information relating to egg color, wingspan, clutch size, courtship display, or some other biologically significant point. If a story line could not accommodate all the information I would have liked, I chose not to overload the story. *But* information relating to courtship, clutch size, nest construction — in fact all the significant facets of raptor life — *are* contained in this book. They may have been thematically apportioned between species, and many have been made the focus of individual essays; first, because the redundancies in closely related species lend themselves to this approach; second, because saying a thing once well is to my mind more fruitful than saying everything over and over again.

While repetition facilitates learning, it works only if readers don't become so numb that they stop turning the pages.

About the facts. Much of the specific information you will find in this book was drawn from the invaluable resource books whose names you will find listed in both the Acknowledgments *and* the Bibliography. These books synthesize the wealth of primary research that has been conducted by a great many raptor biologists. In this book, most hard biological data is unreferenced, not because recognition is unwarranted but because the mechanics strain the nature of a story. In this introduction I would like to make clear that I wish to take no credit for the work of biologists whose fieldwork and insights I both acknowledge and admire. My objective is to disseminate information to a lay audience in the hope that this will result in a broader appreciation for and understanding of birds of prey. Most raptor biologists, at least the ones it is my privilege to know, would understand and approve of this.

About the stories. Some of these, too, are vicarious, rooted in accounts I have read or anecdotes I have heard. Most, however, are drawn from my own experiences in the field. All have at least this much in common. They are, in their essence, supportably true. For example, the male Northern Harriers that you will meet in these pages were displaying just as I have described it. An Osprey pulled under by its prey is a drama I have seen. The altercation between the nesting Peregrine Falcon and a wolverine really happened.

Introduction

The dialogues, the dreams, the expressed emotions, the anthropomorphizing that runs as a thread through many of these stories? These are the constructs of a writer's mind. Though no one can say with certainty that birds of prey do not experience or express life as I have presented it, I cannot (and certainly will not) tell you that they do.

One more facet relating to carriage and language. Traditionally, the term "phase" has been used to describe the light and dark plumage types that are found among some raptor species — most notably the buteos. The term has fallen into disfavor, its usage undermined by the suggestion that "phase" implies transition (something that these plumages are not). In its stead, the term "morph" has been applied.

As one who loves both raptors and language, I eschew the term. "Morph" to my mind, is an ugly-sounding word — ungainly, uncouth, attributes that do not easily come to mind when I conjure the image of a bird of prey. In this book I have chosen instead to use the word "form" to express plumage types. I am not completely happy with it, but at least its enunciation doesn't stick to my adenoids.

The order in which species are presented in this book might interest some (and confound others). Though they are not presented in the phylogenetic order established by the American Ornithologists' Union, there is, nevertheless, rhyme and reason.

The book follows a seasonal progression, beginning with one winter and concluding with the next. The key components of raptor life — territory establishments, courtship, hatching, fledging, migration, and so on — are all laid out along this time line. Individual species were chosen to "star" in these subject areas, and it was this matchup of bird and biological focus that determined the order in which species appear.

Opinions differ as to the number of raptors found in North America. Much of the disparity relates to how broadly or narrowly you define "found." If by "found" you mean "have occurred," then more than forty species have been found — including such extralimital strays as Northern Hobby, *Falco subbuteo*, a Eurasian species; and the Collared Forest Falcon, *Micrastur semitorquatus*, a bird of tropical and subtropical regions.

If by "found" you mean "have occurred and not (or not yet) been accepted by the Committee on Classification and Nomenclature of the American Ornithologists' Union," then the number is trimmed to thirty-seven.

If "found" is construed to mean "resident" or "nesting birds," then the number is further diminished, stopping at the semantic impasses imposed by the White-tailed Eagle — a single, nonbreeding specimen of which resides on the Aleutian island of Attu; and the Aplomado Falcon, a colorful grassland species that has not nested in the United States since 1952 but whose population, thus far nonbreeding, is currently being reestablished in Texas.

The question of standing also involves the New World Vultures, whose traditional ranking among the diurnal raptors has been undermined by recent findings that, though widely accepted, have yet to be officially recognized.

Since *some* decision had to be made concerning the birds to be included in this book, and since *all* decisions are to a degree defensible *and* capricious, I have held to a strict interpretation with regard to residency and a loose one with regard to phylogeny. Accordingly, only birds of prey with established nesting populations are included — a standard that eliminates the Aplomado Falcon. New World Vultures are included — ostensibly because they are still officially recognized as raptors, but also because although they may not be true birds of prey, they are Wind masters in every sense of the word.

Those with a keen sense of jurisprudence will quickly point out that by my stated standards the California Condor, a vulture, has no more right to be counted among the birds in this book than the Aplomado Falcon does. After all, at the time of this writing, there are no nesting wild California Condors either. All adults of propagating age have been placed in protective custody.

But there is, to my mind, a difference between the several years since condors have bred in the wild and the several decades distinguished by the absence of Aplomado Falcons. What's more, the young condors being released in the wild are native stock — direct descendants of birds that were native and did breed in the wild. The Aplomado Falcons being released have no such link to place or past.

Honesty compels me to admit that these are not the real reasons why the condor is included and the falcon is not. Semantic defenses aside, there was something I very much wanted to say in this book, and the condor seemed best suited to play the part. The falcon, thought a wonderful bird and one that I hope will once again have a viable population north of Mexico, would not have allowed me to express my point so well.

There is one final concern I would like to address before letting readers move on to the stories that are the heart of this book. Although widespread education and hunter training courses have greatly reduced the number of raptors that are killed with firearms, this sad practice continues. I would, in rightness and fairness, like to make a distinction between hunters and hawk shooters, because too often hunters are held responsible for crimes committed by anyone with a gun.

Hunters are conservationists who admire and are supportive of the creatures they hunt. Their affinity for the natural world and all it contains is broad and complete.

Hawk shooters are ignorant, selfish bastards who like to kill things.

As a hunter and a conservationist (as well as an environmentalist and a birder), I want all readers to understand this distinction and know that it is true.

— PETE DUNNE
Director, Natural History Information
New Jersey Audubon Society

The Wind Masters

ONE

Falco rusticolus
GYRFALCON

A YELLOW SUN CRESTED the horizon, making the frozen land-scape blaze, stirring momentary interest in the bird perched near the mouth of the grotto. It wasn't prey, and it wasn't *Her,* but it was one of the few times the sun had been seen on Alaska's North Slope since November.

Across the river, on the opposing bluffs, the resident pair of ravens greeted the milestone with acrobatics and irreverent chortles. But the bird in the shallow limestone grotto was as still as the stone that was its perch and as silent as an unspoken thought. Only the head, which moved in measured turns, showed that it was alive and, beneath their dark surface, the eyes that glowed like smoldering peat.

There are no casual observers in the Arctic — not in March. There are only things that hunt and things that are hunted, and this distinction is not absolute because many things that hunt are hunted in turn. But if there were an observer, and if he chanced to see the bird, he would undoubtedly be impressed by its size and color. The

"*There are no casual observers in the Arctic — not in March.*"

The Wind Masters

bird was large, as large as a raven, but white, the color of soot-flecked snow.

If asked to place the bird in the ranks of hunter or hunted, few would hesitate to respond. There was something about the bearing of this bird, something about its visage that was unmistakably raptorial. At close quarters, or through a spotting scope, some of the traits that support this become evident.

The bird's bill, for example, was hooked, well suited for tearing flesh. Its feet were taloned; the hallux, or hind toe, inverted, a refinement tailored for grasping. Above the eye was a flared ridge that both shaded and shielded the eye, and the base of the bill was fleshy, and unfeathered. All these traits are raptor traits and distinguish the bird for what it is — a bird of prey, a creature molded by evolution to capture and kill living prey from or in the air.

Scattered around the bird's perch was more evidence attesting to its predatory prowess. The floor of the cave was nearly a foot deep in regurgitated pellets, the encapsuled and undigested feathers, fur, and bony remains of consumed prey. Most contained the remains of ptarmigan, the bird's dietary mainstay. Others contained the remains of snowshoe hares, voles, lemmings, ground squirrels, and assorted birds.

Not all the pellets had been cast by the cave's present occupant. The powdered remains at the base of the pile had been cast by birds that had occupied this site years before, including some whose heredity was carried in the genes of the cave's present occupant. A few of the larger pellets had been cast by Her, the bird's mate of three years, a robust female the color of silver in starlight.

"*Her*," he thought without realizing he thought it, because as the days had lengthened and the amount of sunlight reaching the earth had increased, the thought had come closer and closer to dominating his mind. "It is time," the bird knew, and this was not a thought. It was knowledge imparted by something more basic and more powerful than thought.

Turning his attention away from the sun, the bird surveyed the world — a rolling expanse of ice, snow, and wind-scoured outcroppings that reached all the way to the Brooks Range. There was little

GYRFALCON

to see, if by "little" you mean "other living things." Alaska's North Slope in winter is a cold, harsh place and it does not suffer living things gladly. Most creatures that can, leave before the first snow. Those that cannot, hibernate or seek shelter beneath the snow, or withdraw into the riverside willow thickets.

Beneath the perched bird, beside a river that would not slip its icy sheath until June, was such a thicket and a large cow moose. On a gravel bar far upstream, two wolves, one gray, one white, scavenged the last frozen remains from what had been the moose's calf.

To some this may seem to be a great deal of activity for a place that feels winter's grip for seven months, total darkness for three. A place where winter temperatures fall to 50 degrees below zero and rarely climb above 0. But human expectations are calibrated to human senses, and the visual acuity of birds of prey operates at levels well above our own. To a human observer, the moose standing motionless in the latticework of willows would have been invisible; the wolves on the gravel bar, beyond the reach of human eyes. But the eyesight of birds of prey is extraordinary, as much as eight times more powerful than human eyesight. It is a thing that humans covet and an asset that serves raptors well.

The bird turned the incredible mechanism of his eyes upon the opposing bluff and the lichen-stained ledge that served as the bird's nest site. The cave in which the bird was perched was a winter roost, and during the nesting season it served as an auxilliary perch — a place to escape, for a time, from the demands of parenting. The ledge was empty.

Once again the bird turned his gaze south, to the sky above the horizon, but it was as empty as the ledge. Casually, the bird turned his head, raised a foot from a skirt of breast feathers, and treated the back of his head to several satisfying licks with a talon. Even the taxon of feather lice, the bane of avian existence, is specific to birds of prey.

An attentive observer watching the bird's hygienic efforts might have noted two traits that linked the bird to a specific family of raptors and distinguished him from the rest. The first was the suggestion of a moustache that scored the side of his face. The other was the

The Wind Masters

size of the bird's feet. They were unusually large. Both these characteristics marked the bird as a falcon, a member of a specialized group within the ranks of raptors.

The thirty-three species of raptors in North America traditionally have been divided into four main groupings: the Cathartidae, the New World Vultures; the Accipitridae, a large and diverse family that includes accipiters, buteos, aquila eagles, harriers, and kites; the Pandionidae, the Osprey; and the Falconidae, the falcons. Recent comparisons of chromosomal proteins, however, have challenged tradition, showing that the vultures share closer kinship with storks than with birds of prey and suggesting that the falcons may merit their own scientific order, one distinct from other birds of prey.

But these groupings, the old and the new, are human constructs imposed upon the world by those who seek to understand it. They have no real bearing on the bird itself, no cause and no effect. And although the specialized traits that distinguished the bird on the cliff, the falcon, from other birds are real, it's not the burden of birds to understand them, only to use them to best advantage.

The feet that distinguished the bird as a falcon were his primary weapons. Drawn into a fist, backed by the weight of a fast-flying bird, they could send birds the size of a Sandhill Crane tumbling to earth; opened into a net of talons, they could rake birds as agile as redpolls from the sky.

The bills of falcons too are specialized. They are notched, shaped like a hand-held can opener, designed to open the cervical vertebrae of prey, letting the life out. And the nostrils are spiked with a bony peg that foils the vacuum that would seal the nares and steal the breath from a fast-flying bird.

There were other falcon-specific traits that only a trained ornithologist would have noted — the projecting keel, typical of fast-flying birds, that serves to anchor the large pectoral muscles that power a falcon's flight; fused vertebrae along the bird's back for structural strength; and a sophisticated tail-muscle assembly that aids in high-speed maneuvers.

But even the most untrained observer could not have failed to note the bird's most celebrated trait, one that has earned the envy of

5

humans since they first turned their faces toward the sky. That is the falcon's powers of flight.

Suddenly, the bird left his perch: one minute he was there, the next he was airborne, cleaving a path toward the opposite cliff on long, broad, tapered wings — falcon wings.

As he flew he called, a harsh *cacking* call that vented annoyance. The raven that was the object of the bird's ire met the challenge with guttural epithets and retreat. The raven had shown too much interest in the falcon's ledge. The falcon had asserted his right of domain.

The falcon did not pursue his neighbor — not this time. The raven had merely tested the falcon's resolve, not challenged it. Ravens and Gyrfalcons, in fact, enjoy a commensal relationship. Their nest-site requirements are similar, and falcons, which do not build their own nests, overcome this deficiency, in part, by appropriating raven nests.

The Gyr circled once in front of and slightly below the ledge, then accelerating, wing tips vibrating, he angled upward, taking a perch upon a lichen-encrusted spire a hundred feet from the ledge. Once more he turned, facing south.

The sun had long since passed below the distant peaks. The shadows that had huddled in snowy hollows had vaulted the hilltops, turning the snow gray. Maybe it was the deepening twilight, or maybe it was the harsh reality of the place. But the white falcon, perched atop his pinnacle of stone, seemed to glow with unreflected light.

For hundreds of years this falcon of the high Arctic has been an object of human admiration. Falconers prize its hunting prowess. Biologists are fascinated by the adaptations that permit it to meet the challenges of life in the Arctic. Birdwatchers the world over are drawn to the mere rumor of a Gyrfalcon. Although the bird's distribution is panboreal and its population healthy and stable, estimated to range between 7,000 and 17,000 pairs, the birds do not wander into the temperate regions favored by our species.

Young birds, without territories, wander more than adults. Adult females may make short migrations to regions offering an abundance of prey, particularly coastal areas, particularly during years when ptarmigan numbers are at the low point in their population cycle.

But adult males are believed to remain on territory even in the Far

North, even in years of prey shortage — though how they accomplish this is unknown.

The ravens had retreated to the shelter of their own ledge. Below, in the willow thicket, the moose stopped browsing and moved to find a place that offered greater shelter. The temperature was falling.

As night deepened, the bird's mind inclined toward the safety and comfort of the roost. But his eyes were held by the horizon.

He *knew* that it was time.

Twice he leaned forward, almost surrendering to the lure of the roost, but did not leave. Once he dropped from the perch, started for the opening, then circled and returned, not to the pinnacle but to the empty ledge, where he walked to the rim of the old raven nest and then back to the edge to resume his vigil. Looking south, he opened his bill and made several low *chupping* sounds that were answered by silence.

Above, the stars were beginning to gather in strength and numbers and the northern lights danced. The moose and the willows were one. The wolves had turned to shadow.

It was nearly dark when the Gyrfalcon finally left the ledge and flew to the roost. In the familiar grotto his feet found perch. He loosened his feathers to increase the insulating distance between himself and the Arctic night.

He sat for a time, letting the tiredness fill his mind. His eyes were closed and sleep only just upon him when, in his sleep, he heard a sound that brought his eyes open and his head erect.

It was like the sound he had made earlier on the ledge but deeper pitched, muted by distance. And even though it was all but night, even though the cliff was deep in shadow, the bird could see a silhouette that had not been there before, a silhouette like his own but with one difference. This one was the color of silver in starlight — though only a falcon could see this.

7

TWO

Accipiter gentilis
NORTHERN GOSHAWK

THE HARE FOUND IT DIFFICULT to see past the pain and the fear; to come to grips with the sudden turn of events and the bird that pinned it to the snow. Dying is an enormous undertaking and quite beyond the cerebral capacity of snowshoe hares, *particularly* since the animal would have so little time to accommodate this new reality. What time it had left in the world would be measured in heartbeats and the volume of blood that was carrying the animal's life onto the snow.

The bird, an adult female Northern Goshawk, had no such difficulties. The bird *was* very familiar with death, was in fact an old confederate. Every day that she lived, something else died, and the bird was now approaching the fifth year of her life.

But if the hare could have gained control of its faculties, it would have seen a large hawk with bright yellow legs and a silver-gray breast, etched with lines so fine they mesmerized. The wings, splayed to both sides for balance, were long, pale, and tapered. Their tips trailed into the snow.

From its unfortunate angle, the hare could not have seen the long, broad, banded tail that branded the bird an accipiter, nor the slate-colored back that marked her as an adult. But it is likely that the animal would have overlooked these things anyway and concentrated instead, as human observers do, on the bird's head and face.

The head was large and broad, cut more in the likeness of a buteo. The bill too was formidably proportioned, gray, dark tipped, and cruelly hooked. Across the face, angling behind her head, the bird wore a charcoal mask, and above the mask, riding the crest of the superciliary ridge, was a pale eyebrow stripe. It gave the bird the suggestion of a frown.

Yet the bird's most arresting feature was undeniably her eyes: large and red and glowing with a ferocious intensity, they were the portals through which the hare could see the new reality that was closing around it. Crimson on the surface, dark beyond measure below, it was a reality that living things do not accept without struggle.

The hare arched its back and kicked out with its hind legs, both the one that was uninjured and the one that was fettered and torn. On its side, bearing the weight of the bird, the animal succeeded only in sending up a spray of blood and snow and drawing upon itself the full killing force of the bird.

As fast as a convulsion the hawk reared back, tightening her grip, sending the talons deeper, searching for the vessels of tissue where an animal's life is housed. The hare flinched in pain and the talons closed again . . . and again . . . and again. . . .

Releasing one foot, leaving the other firmly planted, the bird shifted her grip higher, finding the soft spot just below the ribs. She felt her talons go deep — and the hare stopped struggling. Its heart still beat, its ruined diaphragm still confused. But death had been planted in its body and there was little the animal could do but accommodate it.

The goshawk stared down at the hare, anticipating more struggle, but there was none. Gradually, the bird relaxed her vigilance (though not her grip), and after casting a glance around the clearing to see what threats or challenges the struggle may have drawn, she turned her eyes back upon the hare. The violence was draining from them, but some other expression was replacing it. If a human observer were

9

"As fast as a convulsion the hawk reared back, tightening her grip."

The Wind Masters

asked to put a word to it, he would be tempted, and maybe chilled, to call it satisfaction.

The killing ability of the Northern Goshawk is legendary. As much as the Gyrfalcon's, its power and prowess command awe — although the styles of these two birds could not be less alike.

Gyrfalcons are like skilled boxers, the sky their ring of choice, and the lightning jab, a single knockout punch that sends opponents tumbling, their practiced forte.

Goshawks are street brawlers, tough in the clinches.

Like accipiters everywhere, the Northern Goshawk is a woodland bird, and "woodland" is broadly defined by goshawks. It includes not only the coniferous northern forests, the classic goshawk habitat, but willow thickets in the Arctic North, open pine-oak forests in Mexico's mountains, and deciduous woodlands in the Appalachian east. Vegetation type notwithstanding, goshawks generally prefer mixed habitat — woodlands that offer mature trees, broken understory, open glades, and a surfeit of edge.

The prey favored by North America's largest accipiter is similarly eclectic. Birds and mammals are equally favored, and abundance, more than any other consideration, usually determines what species will win the honor of "prey for the day" (although, like many birds of prey, individual goshawks may specialize, bringing their focus and honed skills to bear on one or two favorite species).

Over much of the Northern Goshawk's range, Ruffed Grouse, Spruce Grouse, snowshoe hares, and red squirrels are the dietary staples. Ptarmigan and ground squirrels are important prey in the Far North; Northwestern Crows, Steller's Jays, and Varied Thrushes in the Pacific Northwest. Waterfowl, to the size of Mallards and American Black Ducks, are actively hunted, and smaller birds and mammals — down to the size of sparrows and chipmunks — are fair game for this master hunter.

Stealth and surprise are the great allies of the Northern Goshawk. If these tactics fall short, then the bird brings a ferocious tenacity to bear. Perch hunting is the rule. From an understory limb or some strategic vantage behind the veil of the forest edge, the hunter sits and watches — for a hare to move from cover . . . a Ruffed Grouse to mount its drumming log . . . a red squirrel that centers too much of its attention on the pinecone in its grasp . . . whatever fortune brings the hunter's way.

Pursuit is direct and silent, the impact terrible. Smaller prey is sometimes killed by sheer shock. Larger prey is bowled over, and as hunter and hunted tumble, across the ground or through the air, the bird brings its feet to bear — and a goshawk's feet are made for killing.

Goshawks have another hunting technique that is well suited to their power and style. If patience proves unproductive, if the sentinels of the woodland are too alert, the birds embark on hunting flights. Sometimes these excursions are short — low, weaving dashes through woodlands, eyes calibrated for panicked movement, wings primed for accelerated pursuit. Sometimes they are more protracted — long, coursing, set-wing glides that navigate the edges of fields and marshes and may cover several hundred yards.

When prey is flushed, the chase is on. A twisting, turning slalom around branches and trees in the case of grouse or hares; a heart-pounding, high-speed, high-stakes race. Cover affords scant protection from a goshawk. The birds careen through branches and, if need be, pursue the creatures they have marked for prey on foot. Goshawks have even been known to follow the tracks of hares through the snow.

The hare that was already prey had been a relatively easy catch. The bird had flown to the edge of the spruce grove an hour before sunset — a place she hunted often and with success. Earlier in the day, she had killed and consumed a Gray Jay. But it had not completely satisfied her food needs. The goshawk weighed 1,100 grams, about average for her sex. The jay, including waste, weighed 75 grams (about 3 ounces). A bird of prey consumes on average approximately 15 percent of its body weight a day, and food demands are more acute in winter. The hawk was facing a night of cold with a fuel deficit to be met. But that was not the only reason she hunted.

After the goshawk had taken her perch, it was less than five minutes before the hare had ambled down one of the well-used trails. The bird waited until the animal paused and raised itself on its hind legs to reach one of the taller, ungnawed branches before launching herself. The rest of the story was written in the snow; the story that was already finished except for one, anticlimactic detail.

The goshawk stared down at the hare, watching the last moments of its life. If her hunger had been greater, if the animal had still struggled, the bird would have footed the hare, killing it quickly. But she did not.

A casual observer might have looked upon the scene and concluded that the bird was winded and catching her breath. A behavioral scientist might suggest that the bird simply lacked for stimulus, that with the twitch of a nerve all the fury of the goshawk's predatory instincts would be released. These things are so.

But there was another reason that the bird waited — one that a scientist, whose focus is limited to things that can be tested and measured, might question and a raptor protectionist, steeped in the coldly rational and apologetic tradition of "predator vs. prey relations," might deny.

The reason the bird paused was to savor her triumph — the way any hunter contemplates an animal brought down by its skill.

The truth, I believe, is that the bird *liked* to kill. It was something projected by the bird's stance; it was something that could be seen, by things that hunt and things that are hunted, in the bird's eyes. And there is nothing, in the acceptance of this, that is incompatible with the prevailing notions that birds of prey are instruments of nature's design; essential elements in the maintenance of a healthy environment.

Nature, like God, works in strange ways. Where was it ever decreed that gods and Northern Goshawks should not enjoy their work?

13

THREE

Buteo brachyurus
SHORT-TAILED HAWK

THE HAWK FLOATED against the clouds, which were still beautiful at midmorning. It would not be until later that the sky would turn dark and heavy with rain. Below her was the gerrymandered mix of forest and grass that is the Everglades. Around her, in the sky, were the other birds of Florida in February — storks, ibis, Anhingas, and, of course, vultures.

Some were turning circles in the moisture-laden air. Others were gliding with some distant ambition in mind. Only one, the hawk, was immobile, poised. Along with supporting field marks — small size, white underparts, dark cheeks, dark trailing edge to the wing, broad subterminal tail band — this disregard for gravity distinguished her as a Short-tailed Hawk, one of North America's consummate yet enigmatic raptors. These hawks have both a light form and a dark form. This bird, a female, was light — in Florida, the less common of the two color types.

The bird held her place in the sky for more than a minute, studying the ragged woodland edge where forest and fields vie for supremacy. Then she began to edge forward on long, tapered wings that turned up at the tips. The intensity she projected upon the earth was mesmerizing. Her control over the air seemed complete.

Very few birds of prey are able to "kite," that is, hold their place in the sky without flapping. These include the very common and widespread Red-tailed Hawk, the western Ferruginous Hawk, the White-tailed Hawk of coastal Texas, and the White-tailed Kite. All but the last of these are buteos like the Short-tailed Hawk. And while all birds of prey are celebrated for their powers of flight, few, and maybe none, are so adept at stop-and-go control as the Short-tailed Hawk. These tropical buteos seem to defy gravity in a way no other birds of prey can do. Their primary prey is other birds, just as it is with falcons and accipiters, and this is a difficult specialization for a bird of prey because the flying skills of prey nullify a raptor's greatest strategic advantage.

The bird paused again, stopping her progress with no more apparent difficulty than a car braking in traffic. Only a slight lengthening of the wing betrayed the secret hidden beneath the magician's cape. By changing the set of her wing, the bird was able both to manipulate the updraft coming off the woodland edge and to capitalize on the rush of air passing over her wing to neutralize gravity. It was a neat trick, one no falcon or accipiter has ever mastered.

There are in North America one harrier species, the Osprey, two eagle species, three accipiters, five falcons, five kites . . . and twelve buteos (almost as many buteo species as all other raptor species combined). The members of this large tribe share gross anatomical similarities: medium to large size; long, broad wings; generally short tails. But within their ranks they distinguish themselves by size, shape, anatomical refinement, plumages, nesting requirements, habitat preferences, prey, and geographic distribution.

Charles Darwin, in his *Origin of Species,* was the first to explain the principles governing evolution (the mechanism) and species development, evolution's product. His most famous edict relates to competing organisms — how the fitter will survive and reproduce and the less fit perish. Between two members of a species, whose

"*She began to edge forward on long, tapered wings that turned up at the tips.*"

The Wind Masters

relationship may be that of genetic competitors, and between species for which the relationship is that of predator and prey, Darwin's maxim holds true.

But the heart of Darwin's insight had to do with *accommodation*, not *conflict*. What Darwin observed was that modifications in anatomical design (and behavior that applies these advantages to best effect) were the mechanism that permitted species to coexist. It is precisely these modifications, which we call specialized traits, that enable birds of prey in general, and even closely allied species such as the buteos, to flourish across North America. Specialization gives different species the latitude to occupy habitat unsuited for another. Specialization also permits birds of prey to coexist in the very same habitat by utilizing different prey or by securing prey by different means.

While the various groups of Wind Masters enjoy specializations that set them apart, it is the very breadth of specialization found among the buteos that makes them special. No other group excels in so many ways, and no environmental niche seems overlooked by this versatile assemblage.

In her patented stop-and-go manner the Short-tailed Hawk navigated the length of the ecotone, then, wheeling around, she returned to the projecting tip of the woodland to work the edge again. For all the world, it seemed as if the bird were playing the game "red light, green light."

Suddenly the bird, who was already motionless, became stock still. She didn't tense. In fact, quite the opposite. It was more nearly a calm, and it defined the difference between immobility and utter immobility.

On the ground, 600 feet below, a red light had flashed — the flared epaulets of a Red-winged Blackbird. The bird was atop a shrub displaying for a female and, though he was not stupid, was unaware of the hawk's presence.

He was unaware because the hawk was very high, at the edge of the songbird's threshold of concern. Short-tailed Hawks exploit this edge. He was also unaware because the hawk was motionless against the sky. It is motion that catches an eye; immobility defeats it. Set against the friendly backdrop of clouds, the white-bellied hawk, with wings trailed in shadow, was almost invisible.

The hawk began to descend. She didn't fall. She parachuted, slowly, like a person falling in a dream. No falcon and no accipiter could have executed such a maneuver — and very few buteos. When she'd narrowed the distance by 150 feet she stopped . . . and *plunged*. Headfirst. Wings back. Like a taloned teardrop. A falcon making such a stoop would have been proud; an accipiter, reckless. Thirty feet above her target she extended her talons and opened her wings to check her descent. The blackbird, intent upon a female, didn't realize death was in the air until talons ferried him across the threshold. The hawk was back in the air almost immediately, heading for a favorite perch.

From the opposing wall of trees, another bird took wing and began circling — a buteo, similar to the Short-tailed Hawk but different. The newcomer was ruddy breasted where the Short-tailed was white. Its tail was narrowly banded, black and white; not gray and broadly tipped like the Short-tailed Hawk's. Its wings were rounded, like a forest hawk's, not tapered like those of a hawk that hunts in open air. Its rust-colored shoulders identified it as an adult Red-shouldered Hawk. Its cries expressed a high level of agitation.

The Red-shouldered Hawk and the Short-tailed Hawk were uneasy neighbors — in fact, their hunting ranges and their prey partially overlapped (hence the irritation). But confrontation between the two birds was kept to a minimum and coexistence dominated their relationship. This was how the birds kept the peace.

First, the competition for prey was tangential, not complete. Whereas the Short-tailed Hawk is almost exclusively a bird-catching raptor, the Red-shouldered is eclectic. Birds constitute a very small percentage of the Red-shouldered's diet — less than 10 percent during the nesting season.

Second, the hunting styles of the birds differ. The Red-shouldered is a perch hunter, like an accipiter. The Short-tailed is an aerial attack artist, more like a falcon.

Third, and closely related to the respective hunting styles, Red-shouldered Hawks tend to hunt early and late in the day. Short-tailed Hawks, who make concerted use of thermal lift, are most active when thermals are active — a period that mantles the middle of the day.

Fourth, in Florida Red-shouldered Hawks begin nesting by the middle of February. Short-tailed Hawks are commonly four to six weeks behind and may not begin incubating eggs until late March or early April. This temporal buffer ensures that the nestling periods, when competition for prey is most acute, do not overlap.

There are two other factors that serve to minimize conflict between these species in particular (and all raptor species in general). The first is distribution. Both species occupy vast geographic areas. The Red-shouldered Hawk is found over most of eastern North America (with a separate population localized to California). The Short-tailed Hawk is found across much of the Neotropics — southern Mexico, Central America, and most of South America. The only place where Short-tailed Hawks and Red-shouldered Hawks overlap is the Florida peninsula — the total expanse of the Short-tailed Hawk's range in North America (one of the most limited ranges of any North American bird of prey).

The other factor helping to minimize interspecific interactions is population. Though in Florida the Red-shouldered Hawk is widespread and common, Short-tailed Hawks are few. The best estimates place the population at 500 adults, or 250 pairs.

In winter, virtually the entire population retreats to Lake Okeechobee and south. In late January or early February, nonresidents depart and disperse north to breed — some only as far as Brevard County on the Atlantic side of the peninsula; Gulf-side nesters stretch almost the length of the peninsula. In a stand of mature trees, close to the edge habitat that Short-taileds like to hunt, they will build a nest and attempt to raise two young. With luck, and all of an adult Short-tailed Hawk's skill, statistics say on average they may succeed in raising one youngster to fledging.

The female Short-tailed ignored the Red-shouldered Hawk and took a perch at the edge of the woodlands and proceeded to feed, tearing pieces of Red-winged Blackbird free with her bill and offering a tribute of feathers to her ally the wind. Overhead, the clouds had lost their morning friendliness and were beginning to take on the dark demeanor of clouds that mean to rain.

Birds were hurrying to conclude their business before the deluge — except for one dark-bellied hawk who seemed intent on facing down the wind. It was the Short-tailed's mate, as specialized a hunter as she. Against the backdrop of the fast-darkening sky, he was almost invisible.

DAS

FOUR

Catharteſ aura
TURKEY VULTURE

THE LATE-MODEL STATION WAGON, its headlights still blazing, weaved a path up Patriot Avenue and turned left onto Muster
Drive. A gloved hand directed plastic-wrapped missiles at the Clairs'
driveway and the Moodys' porch before cresting the hill and disappearing from view.

The vulture watched in silence.

The upstairs bathroom light winked on at the Goldblooms' split-
level ranch a split second before the kitchen light went off at the
Mayhoods', while two doors down at the Patels', the clock radio
abruptly played the last line of a country western song that was
pushing the top of the charts, then switched over to a string of
Presidents' Day sales announcements.

The vulture noted all these things also — although it didn't care,
one way or the other, about music and it didn't need any appliances.

As a trio of shadow-colored deer nibbled at Bob O'Mally's box-
wood hedge, and Lois Baker opened the back door to let their three-

year-old schnauzer, Charlemagne, attend to business, the sound of the 6:22 pulling into the Bernardsville station lofted out of the valley. Even from the vulture's vantage of its hilltop perch, close to the trunk of a grand old tulip poplar, the train was not visible. Nor was the commuter traffic building on nearby I-287. Nor was the sun that was, at that moment, painting a beautiful dawn over the hills of northern New Jersey.

The vulture took it all in with piety and silence. It was part of the morning ritual.

From a nearby branch, another vulture announced its departure from the roost with a series of noisy flaps. The air was cold and windless; the bird's flight, clumsy and effortful. Once it had cleared the trees, the bird set its wings and began a shaky downhill glide that would carry it over the town. It was a young bird, a subadult, reared the previous summer. Its foraging success had been poor of late and the night had been cold, energy sapping. The bird's energy needs were severe.

Solemnly, the other vulture, an adult, watched it go but made no move to follow. If the young bird had come in to the roost the night before with a full crop, it might have been different. Then the adult might have followed, because had the youngster been privy to a food source, such as a road-killed deer, it might be something another vulture could exploit.

But the youngster had no such resource and, as it was, the adult was merely hungry, not starving. Survival, it knew from experience, was as much a matter of conserving energy as finding food. Like most of the other vultures in the roost, it would wait for easier flying conditions before taking to the sky.

There were nearly fifty birds scattered across the hilltop roost, most of them Turkey Vultures. It was light enough now to distinguish the several, smaller Black Vultures in their midst; light enough to see the ivory-colored bills of adult Turkey Vultures and to suspect, if not actually *see*, that the heads might be some other color than gray.

As vulture roosts go, this one was not large — there are records of roosts in which more than a thousand birds have gathered. And there was certainly enough perch space in the trees of the hilltop to support

TURKEY VULTURE

additional birds. But space was not the limiting factor. The problem was food — and the amount of time it takes to secure it.

With little to do until it was time to leave the roost, the bird remained quiescent, its feet planted firmly on the branch, its unfeathered head drawn down into a protective ruff of feathers. Some people, including some of the neighborhood residents, consider vultures vile. It would take a generous heart to call them attractive, but the birds are not without character.

Turkey Vultures are large birds, cloaked in loose-fitting brown/black feathers that make them look shaggy and unkempt when perched. The legs and feet are sturdy and unfeathered; the color, where not coated in excrement, is cherub pink.

The face is ruddy and wrinkled and topped with an embarrassingly sparse cap of feathers. Its expression is uncommonly serene and there is something about the face that suggests infinite patience. Huddled on their perches, wrapped in shabby vestments, the birds look like a group of balding monks gathered in prayer.

Appearance, of course, is not the only root of the vulture's popular disfavor. Vultures are primarily carrion eaters, their feet unsuited for grasping, killing, or carrying prey, as the feet of true raptors are. Although Turkey Vultures do occasionally capture and kill small prey and have been known to consume a rotting pumpkin now and again, the flesh of freshly killed animals is their main fare. Interestingly enough, the creature largely responsible for providing the vulture with food is the same one who regards it with loathing and contempt.

Blake Mayhood's garage door opened, disgorging a beige-colored sedan, which backed onto the street and, accelerating quickly, headed in the direction of the interstate. The deer that stepped in front of the car owed its life to antilock brakes and new radial tires (and Blake's quick reactions). The anguished squeal of tires brought the heads up on a dozen vultures.

Deer populations and suburban sprawl throughout the Northeast have more or less kept pace — offering possibilities to scavenging birds such as the vulture that did not exist half a century ago. Five-acre zoning, a surfeit of edge habitat, and protection from hunting that populated areas confer have resulted in an exponential growth in deer populations in many areas.

But the two elements that made suburban development possible, the automobile and the highway system, are also, to a large degree, the factors that allowed Turkey Vultures to expand their breeding and wintering range in the Northeast. The highways are the killing fields; the automobile is the top predator and, from a vulture's standpoint, an ideal predator, too. Not only does the creature kill deer and other wildlife regularly, it doesn't eat what it kills.

Blake Mayhood waited until the last deer had crossed the road before continuing. The vulture watched the deer disappear into the trees and then it started to preen. Birds of prey spend most of their lives perched, and much of this time is spent preening.

A reliable source of food serves for nothing unless it can be exploited, and vultures trying to expand their wintering range north face a problem that those birds which withdraw to southern areas during the winter do not — time. The farther north you travel, the less daylight you find and the less time vultures have for foraging. As soaring birds that rely upon updrafts or thermals for lift, their movements are dependent upon wind and insolation. On a windless morning, birds must wait until the sun is high enough to send thermals aloft. In midwinter, thermal production may last no more than four hours — a narrow window of opportunity. On a cloudy day, there may be no opportunity to forage at all.

As much as prey availability, foraging time is the consideration that forces Turkey Vultures nesting in southern Canada and northern states to withdraw in winter — even if only for a few short weeks. Some Turkey Vultures are still heading south in December; a few are already returning north by the first week in February. In the East, wintering birds are heavily concentrated in Florida. In the West, California and Arizona are winter strongholds, but many thousands of Turkey Vultures withdraw to Central and South America.

The first rays of the sun touched the hilltop, falling upon the perch-bound birds. As if in some practiced dance movement, most of the birds responded by raising their wings in a gesture of greeting. It looked like a benediction. It looked like a hundred silver flowers blooming or mirrors flashing in the sun. Some of the birds turned on their perches, exposing their dark backs to the light, absorbing more of the radiant warmth. Vulture body temperature drops 10 degrees centigrade while the birds are sleeping — an energy-saving adaptation on a par with suburban homeowners turning down their thermostats at night. Shivering brings body temperatures back up to daytime standards — but sunlight does it almost as quickly and without any waste of energy.

The sunlight was one reason why the birds had chosen the hilltop stand of trees for their roost. A roost in the valley would have kept them in the shadows longer — and this would have shaved precious minutes off the short winter day. Elevation also favored the roost site. Even without the sun, even without a wind, the birds could still leave their perches and glide toward the valley and the interstate highway below.

More birds, mostly immatures, began leaving the roost, cashing in on the elevation they'd bought with the previous day's thermals, the thermals that they'd used to gain the elevated hilltop. The birds launched themselves with a noisy flurry of wing beats, then set their six-foot wings in a glide. Most flew directly away from the roost. Only a few tried to tease lift out of the air by soaring over the roost, and these failed. The dispersal of birds was purely random — although each bird was guided by purpose and design.

The adult who had watched the neighborhood awaken remained until half the roost had departed — and until it noted the first circling birds that were gaining elevation. Throughout the day, it would be watching those other birds, using their eyes and their fortune, hoping to take advantage of whatever fortune fell their way.

It left its perch with a push, flapped three times, and then started a long, flat-winged glide that would carry it over the rooftops of the community. Nearby was the parking lot of a daycare center, set at a slight angle, the asphalt top of which quickly absorbed and radiated heat on cold mornings.

The bird did not hurry. A carrion feeder has no particular need for hurry. It did not flap. Instead it used the angle of its glide to best advantage — balancing the forces of friction and gravity until it

The Wind Masters

"It looked like a hundred silver flowers blooming."

TURKEY VULTURE

gained the edge of the lot and felt itself buoyed aloft. For the first time since it left the perch, it turned and pirouetted in the sky and its wings flashed silver.

On its perch, there is something humble about the Turkey Vulture. But in the air, it is a thing transformed, a creature perfectly attuned to its environment.

Haliaeetus leucocephalus

BALD EAGLE

THE SUN LIFTED above the valley's rim, falling upon the low-lying fog that rose to greet it. As the mist curled skyward, disappearing into the cold, thin February air, it offered momentary glimpses of packed ice, dark water, and a wall of great-limbed trees.

Below the upper rapids, several shadowed forms stood upon the ice. They might have passed for boulders, but the stately white-headed bird, perched in his customary place on the limb of the sycamore, knew better. The opaque phantoms were Bald Eagles, like himself, and they were gathered around the remains of a deer carcass imbedded in the ice. The eagle knew this, too.

The eagle that was perched, an adult, was not particularly interested in the deer, although he had fed from the carcass once or twice and might again — if other options failed or if one of the phantoms managed to secure a portion worth appropriating. Piracy is an essential tool of eagle statecraft, and adult eagles are not shy about relieving less experienced birds of contestable goods.

The fact was, the adult was not partial to frozen deer. What little

"*The stately white-headed bird, perched in his customary place on the limb of the sycamore, knew better.*"

The Wind Masters

interest he initially had had in the carcass had diminished as the carcass had diminished and as the number of birds fighting for scraps had increased.

"Too much effort now, for too little gain," the eagle concluded, and his proximity to the carcass was pure serendipity — which is *not* to say that the bird's choice of perches was capricious. Quite the contrary. There were very definite reasons why the bird, when he had left the roost at dawn, had made for this spot. These reasons had much to do with an eagle and its needs.

The favored tree was tall, nearly seventy feet tall; the limb that served as the bird's perch was horizontal and sturdy. Bald Eagles are large birds, weighing from eight to fifteen pounds. They like their trees large and their perches strong, unencumbered by branches that might get in the way of their seven-foot wingspan.

The tree stood on a sweeping bend in the river, offering an unobstructed view north and south for a span of some four miles. What's more, the stretch of river boasted two large sets of rapids that never froze, and this was key. Fish is the staple food of Bald Eagles, and in parts of the bird's range, a range that covers most of North America south of the tree line and north of Mexico, fish may constitute 80 percent of an eagle's diet.

Fish, of course, do not need open water in order to survive. Eagles, however, require open water to secure them. This is why inland-nesting northern eagles retreat to the coasts or to fast-moving rivers when winter closes its fist over the land. This is why the adult perched in the sycamore favored this particular stretch of river.

There was an assortment of fish in the river — walleye, pike, trout, and small-mouth bass, although eagles generally favor slower, easier-to-capture fish such as suckers and catfish, the so-called rough fish. And the bird is wonderfully adept at locating and capturing live fish — either by actively cruising the river or by using well-situated perches to its best strategic advantage.

But the adult eagle (who was more partial to fish than deer) much preferred to claim winter-killed fish carried by the river to catching live fish. Dead fish are easier to secure than live ones, which is more energy efficient, and patience is all that is required. Like piracy, "wait and see" is an important tenet of eagle statecraft, too.

In addition to fish, the river offered another epicurean enticement, an assortment of waterfowl. These included Common Goldeneye, Common Merganser, and a large flock of puddle ducks concentrated in the open water downriver. Waterfowl are even more difficult to catch than most fish, so here again eagles play the role of opportunist. Ducks that are sick, injured, or winter-weakened are easier to secure than healthy birds and as prey are favored. This fundamental principle of predation is not absolute, but it is universal, and eagles are confirmed practitioners.

The eagle's perch held one more strategic advantage, one that might not be immediately apparent to a human observer (and this is very much the point). The perch was isolated — far from roads, far from houses, buffered by open water and unbroken woodlands far from human disturbance. Eagles require privacy, and this demand extends equally to hunting perches, to roosts, and to the trees that bear their great stick nests.

Among themselves, eagles may be tolerant or intolerant of intrusion, and their prevailing attitude depends largely upon the season and the availability of food. Where food is found in abundance, such as the winter salmon concentrations in Alaska's Chilkat River, thousands of eagles may gather to feed. More commonly, wintering eagles roost communally, but space themselves during the day. In this way, they both avoid competition and increase the chance that some member of the group may locate a food source that other birds can exploit (diplomatically or otherwise).

If food had been harder to come by, if the adult eagle had been more energy stressed, he might have been less tolerant of the eagles worrying the remains of the carcass on the ice. But the bird was in excellent condition, and so long as the other birds did not infringe upon the adult bird's river rights, he was prepared to be magnanimous (although "guardedly indifferent" probably comes closer to describing the bird's true feelings).

The other adult eagle, perched among the hemlocks on the far side of the river, was different. Her perch offered the same vantage as his. Her interest focused upon the same stretch of water. Her skills were as refined and this made her a fiercer competitor.

But if anything, the sycamore-perched male was even more tolerant of the female than of the dark-headed shadows on the ice. First, because as a female she was considerably larger than he was. Second, she was not just any adult eagle. She was the bird's mate.

Bald Eagles pair for life, and although northern birds that leave their territories in winter may not necessarily remain together after the breeding season, these two did. They migrated south as a pair in November and would return together in March. They shared the same twenty-mile stretch of river in the winter.

The pair's nest was in a tall, lightning-topped white pine on a lake in northern New Brunswick, Canada. He was sixteen years old; she, his second mate, was nine. They had paired four seasons earlier in her first year as a full adult, and they had successfully raised six young, two per season, which is average for the species.

That the female was an adult Bald Eagle was easy to see. Both her head and tail were brilliant white, the body brown onto black. Except for the talons, which were black, the bird's eyes, bill, legs, and feet were bright yellow. This was the bird whose bearing had inspired the Founding Fathers to adopt it as the symbol of their nation, and its image is familiar to every American.

The birds on the ice did not inspire instant recognition. In fact, only one of the four resembled the adults, and none bore likeness to any of the others although all were Bald Eagles. They were just eagles of different ages and in different stages of plumage development.

All birds of prey undergo a series of developmental plumage changes as they mature, beginning with two successive coverings of natal down that they acquire soon after hatching. Before birds leave the nest, these downy coverings are replaced by a covering of juvenile feathers that are structurally similar to the feathers worn by adult birds — though in most cases they differ in color, even in length. The flight and tail feathers of young raptors, for example, are commonly longer than those of adults, a structural advantage that partially compensates for a younger bird's lack of aerial finesse.

Most birds of prey undergo one or two full feather replacements, or molts, en route to their adult plumage. But eagles are longer-lived than most birds, maturation and plumage development more protracted. Young Bald Eagles pass through as many as six plumage phases en route to the covering they will wear as adults — the so-called Basic V plumage. This maturation process takes between four and five years and is completed in the fifth or sixth calendar year of the bird's life.

Among the birds on the ice, the largest was also the youngest. She was a second-year bird, a distinction she acquired on January 1 by surviving to reach her second calendar year. But except for several pale, new feathers on her face and neck, her plumage was still that of a juvenile eagle, a first-year bird right out of the nest.

She was overall dark, the color of dark chocolate. Only the dirty white base of her tail and pale underwing linings hidden beneath her folded wings marred this uniformity. Her bill was dark gray, her eyes brown, and her feet, poised atop the remains of the deer, were dull yellow.

Flanking the young female were two battered-looking birds sporting large amounts of dirty white feathers over much of their bodies. One of these birds was dark headed, though paler than the juvenile, and white bellied. He was distinguished by a white triangle of feathers on his back and a dark band of feathers across the chest that suggested a bib.

The other eagle was even more unkempt. His head and body were liberally spattered with white, the belly blotched, the bib shabby and shopworn — as in fact were many of the bird's feathers; broken and worn. Next to the bird's piebald appearance, his most distinctive mark was a dark line passing through the eye and angling down the neck. It gave the bird a distinctly Osprey-like appearance.

The bibbed bird was one year older than the young, dark female, a bird in Basic I plumage. The Osprey look-alike was two years older than the female in what is called Basic II plumage. Both were noticeably smaller than the female, clearly males, and though both were respectful of the female's size, neither was particularly impressed by her claim to the carcass. In fact . . .

The bibbed male approached the carcass with a straight-legged gait, ignoring the female, who reared to her full threatening height. When it became clear that intimidation would not work, the male

The Wind Masters

changed tactics. He began working around the female, trying to find a portion of the carcass beyond her reach. The female responded to the challenge by charging the intruder and chasing him across the ice, leaving the carcass unguarded. Before the female could return, the masked bird absconded with a piece of viscera that, to his mind, more than compensated for the vengeful tail chase that followed. Needless to say, during the harangue, the bibbed eagle (who now had the unguarded carcass to himself) quite agreed.

Opportunism (like patience, like piracy) is a tool of eagle statecraft.

The fourth bird on the ice, the near adult, stayed out of the fray. Her focus was directed toward a section of hindquarter that a coyote had moved away from the main carcass. She was calmly working the last frozen strips free with a bright yellow bill — identical to those of the adults. This female was a fifth-year bird, nearly four years old, and only a few dark feathers on her white head and some dark mottling on the tips of her otherwise sparkling white tail distinguished her from the adults.

Like the adults, she too had learned the value of not wasting energy. Energy wasted was energy that would have to be replaced. Despite their longevity, despite their protracted juvenile apprenticeship, most eagles never ascend to this important level of eagle statecraft. More than 90 percent of the eagles that fledge in a given season do not survive to adulthood, and nearly 60 percent of these die during their first year. Starvation is a young eagle's greatest adversary.

The adult male, who was a very accomplished statesman (and who really did not care for deer at all), was nevertheless hungry. The sun that had unveiled the river had disclosed nothing by way of an easy meal. Its rays, falling fully upon the earthen bank of the opposite shore, sent heat waves rippling through the air, cast bits of the previous year's grass spiraling aloft, and brought the eagle to a decision. With a glance at the perched female who he knew would be watching, the adult leaned forward, opened his wings, left the perch, and started quartering downriver.

The bird's wing beats were languid but powerful, steady and direct. He passed low over a flock of mergansers, who honored him by diving; then, reaching the far shore, he spread his wings, banked easily, and began turning lazy circles in the air, riding the first strong thermal of the day aloft.

Trading thermal for thermal, climbing steadily, the bird moved slowly downriver — something that should have given discerning eyes cause to ponder. Had altitude been the bird's only objective, the southerly winds would have carried the eagle upriver. Clearly the bird was guided by some purpose, because an eagle does not fly against the wind for no gain.

When the bird's white head and tail were invisible against the sky, when his wings were no more than a dash that could be lost to the blink of an eye, the dash folded in upon itself and began to glide toward the horizon.

It seemed that the bird was not moving very fast, but this was illusionary, a trick played by distance. The bird was moving *very* fast, and as he fell, trading altitude for speed, he moved faster. The ducks who had watched the eagle's climb, and who had seen it all before, moved nervously toward the center of the river.

At first gravity alone drew the missile on, but before the distance to the horizon had been halved, the bird added the power of his wings and his speed increased. The eagle did not fly directly toward the flock. He chose a circumspect course, parallel to the river, that left room for doubt to grow, indecision to flourish, and hesitation to bind the ducks to the river. Only a handful were bold enough to surrender their anonymity and take their chances in the air. The rest vied to be among the birds in the very center of the mass, and this feathered fusion lasted until the eagle abruptly veered in their direction, making their folly manifest.

The panicked flock rose in a blizzard of wings and headed directly downriver. The eagle, who had anticipated this, did not even need to alter his course. He merely quickened his wing beats and increased his speed.

Ducks are fast fliers, capable of sustained speeds of 60 to 70 miles per hour, and ducks flying for their lives may be faster still. But the eagle caught the fleeing birds easily, reached out and down with open talons, and scooped up a young mallard flying in the rear ranks just as easily as he would have lifted a fish from the surface of a lake.

The Wind Masters

It had been this way all winter. The adult eagle (who truly disliked frozen deer) was *very* partial to duck and very adept at catching them. He managed to secure a bird nearly every day, but despite this, the ducks had never developed any strategy more creative than flying straight away. Ducks are not accounted among the planet's great strategists.

His brunch secure, the eagle wasted no more time or energy. He landed on the ice, pinned the duck with one foot, and, stripping mouthfuls of feathers from the bird's breast, he started to feed. He managed to bolt down most of one breast and part of the other before the female reached the scene, forcing the male to surrender the rest of the bird to her.

It had been this way all winter, too.

It wasn't exactly piracy. It was more akin to deference or maybe tribute. It wasn't a matter of statecraft either. More nearly, it was a synergistic arrangement, one that was detrimental to neither bird and ultimately beneficial to both. In a few short weeks, the eagles would begin their journey north. Upon reaching their nest, beside a still frozen lake, both would be in prime condition, ready to begin the serious business of procreation — as biological objectives go, one that is second only to survival.

The adult male did not return to the sycamore. It took another, closer perch instead — another one he favored. From his branch, close to the trunk of the white pine, the bird could watch the rapids, watch the female, and study the fortune of the young birds upriver.

It was a good perch. Out of the wind. Out of view. Isolated from any disturbance. A good place to sit and do nothing — the privilege of accomplishment and a tribute to his skills as an eagle and a statesman.

Falco columbarius
MERLIN

IN THE BEGINNING was the bird, who was hungry, and who having made his decision left his usual perch on the old duck blind and started off across the marsh, toward the flats. Perched, the hawk looked like part of the blind itself — a protruding stave rising above a crossbar, weathered to a blue luster by the sun and salt-laden winds. But in flight the bird became precisely what it was: a compact, wedge-winged falcon, hardly larger than a robin, that flew upon wings that moved as precisely and as passionlessly as pistons.

Aside from the directness, there was nothing about the bird or his flight that could be considered threatening — at least to innocent eyes. The Black-bellied Plover, who whistled a plaintive warning, and the several Least Sandpipers, who beat a zigzag retreat, might have been overreacting.

But after five months of interaction between the predator and his prey, there were no innocents left on the flats. The anxiety of the birds was well founded. It was late afternoon, feeding time for Merlins. Some birds of prey are flexible in their feeding habits, but Merlins are as punctual as they are direct in their flight. Twice each day — once in the morning, once in the afternoon — something wearing feathers and lying within reach of a Merlin's wings dies.

In the summer, on the prairies, the tundra, the taiga, or the burn opened northern forests where Merlins breed, prey might be passerine species — a sparrow, a Horned Lark; a warbler or a thrush. Some small mammals are also taken. Insects figure in the bird's diet, especially the diet of birds newly on the wing. But for much of the year, in the coastal areas where many Merlins choose to winter, prey is usually small- to medium-size shorebirds.

So to the discerning eyes of the shorebirds spread across the Pascagoula marshes of southeastern Mississippi, there was nothing ambivalent in the falcon's flight. One tenth of their number, nearly three hundred birds, had paid for the knowledge that now sustained the rest. The crux of this hard-bought piece of intelligence was this: When the one who watches the flock comes, he kills. It was intelligence worth knowing.

The falcon, an adult, was in excellent health and in the prime of his life. He knew his flock as well as the flock knew him, and both knew that the falcon's supremacy was not complete. In fair chase, discounting all circumstantial advantages, the skills of the healthy birds in the flock were a fair match for the falcon's attack. But the falcon and his flock were privy to another shared understanding. Both knew that somewhere within the honed ranks of shorebirds there was one whose ability to survive was inferior to that of the rest. This one would be prey.

Predation, as it is regarded and defined, is an interaction between living things in which one partner victimizes the other for its own gain. This gain is one-sided, meaning that only the victimizing partner benefits. The loss is also one-sided, limited to the prey, and it is utter.

To adhere strictly to this definition, the interaction between the falcon and the bird that would ultimately become prey does qualify as predation. The falcon lives; the shorebird dies. Very clean. Very simple.

But what about the relation of the falcon to the flock? Is it accurate

37

or even fair to characterize the relationship between the Merlin and the wintering shorebirds as predatory?

The loss of a single bird is not utter because most of the birds are unscathed. Neither is the gain one-sided. The falcon, by culling a slower, and possibly disease-weakened, member out of the flock may by his action be preventing the spread of infectious disease. This is certainly beneficial to the mass of birds in the flock, not at all one-sided.

Consider, too, that when the falcon removes a member of the flock whose genetic dowry is impoverished before that bird can pass its weakness on to another generation, the species as a whole benefits. And if a bird is injured, so injured that it cannot return to the Arctic to breed but not so infirm that it cannot compete with healthy, genetically viable members of the flock for food, its removal from the flock benefits those who will survive and pass their inheritance on. The removal of such a bird favors the species. It hardly meets the criterion established for a one-sided act of predation.

In the relation between the falcon and his flock, does the predator *really* benefit unconditionally? Is the capture and killing of a member of the flock all pure predatory gain without mitigating loss? After all, the falcon is forced to expend energy to capture prey. Even with experienced, adult Merlins, only about one out of five attempts to capture prey is successful.

What's more, every hunt, successful or not, is a lesson in survival for the shorebirds and a strategic loss for the falcon. Survivors get to sharpen their survival skills and study the strategic strengths (and weaknesses) of the falcon. This is certainly beneficial if a bird wants to survive. Prey species learn how to survive as surely as predators learn to kill. And over the course of a winter, the skill levels of hunters and hunted move toward a higher level of perfection and closer together as the margin between predator and prey draws fine.

So it might be argued that the relation between the falcon and his flock is not strictly speaking predatory. It is, more nearly, an interaction in which both gain something, both lose something, and both move each other toward a higher level of perfection where the end, which is survival, is one. In a very real sense, the contest is not between the falcon and its prey at all. The contest is between the

The Wind Masters

members of the flock because it is the flock, after all, that shrugs off its weakest member, condemning it. The falcon, for all his predatory finesse, was no more than the instrument of the flock's will.

Neither the falcon who was flying across the flats nor the shore-birds fanning out before him in panicked waves were dealing in philosophical abstractions. They were dealing in survival. The falcon was cruise hunting, moving in a low, fast, horizontal flight hardly two feet off the flats. Cruise hunting and perch hunting were the two techniques the bird used most commonly and most successfully.

Home for the bird was the edge of a spruce forest bog in northern Ontario. The nest, in typical falcon fashion, was a preowned nest. Over much of the Merlin's range this means crow, raven, or magpie. Only in the northern extremes of the bird's North American range, when the taiga gives way to tundra, do Merlins nest on the ground, as this bird of Holarctic distribution typically does over much of Eurasia.

Young Merlins, except for their brown rumps, look much like their gray-rumped mother — brown above, cinnamon streaked be-low, brandishing the pale facial slash, or "moustache mark," that distinguishes their irascible kind. The adult male, however, looks a bird apart — blue above, rose spattered below. Of the three North American races of Merlin — the taiga race, the prairie race, and the "black" race of the Pacific Northwest — the prairie race is by far the fairest. But the late-afternoon light spreading across the flat made the taiga bird's upper parts gleam like an oiled sword blade in the sun.

The falcon was making no effort to conceal himself. He simply moved down the beach, sending up flocks before him. Some rose and flared in coordinated groups that lit in the falcon's wake.

These the falcon ignored.

Some charged offshore in rising clusters or spreading sheets or writhed like banners that broke like wind-whipped smoke into small groups.

These seemed to interest the bird more.

The Merlin bided his time, directing his course toward one group, then another — testing the birds' reaction, searching for something that would catch his eye or betray a subtle weakness cloaked beneath the anonymity of the flock.

"*The falcon was making no effort to conceal himself. He simply moved down the beach, sending up flocks before him.*"

40

The Wind Masters

Abruptly the Merlin veered offshore, moving toward a splinter flock of Dunlins that had coalesced, disintegrated, and coalesced again. Something in their maneuvering suggested a flaw, a weakness. In a group, weakness almost always comes down to an individual. Once pressed, it doesn't take long for the flaw to show in an otherwise seemingly seamless garment.

The flock, realizing it was the target, rolled like a wave, flattened out above the water, and accelerated. One flanking bird fell behind, and when the flock cut left, the laggard, a first-winter bird, didn't even turn.

It was almost as if the flock itself had shrugged off the struggling bird. Or maybe the Dunlin, knowing that its anonymity had been compromised, realized that the flock held no more advantage.

For whatever reason, for all that really mattered, now it came down to two birds. One predator; one prey. Both enjoined in a drama almost as old as life itself. The falcon accelerated and the Dunlin tried to match him . . . but failed. The falcon closed, climbed above the shorebird, accelerated, and missed as the Dunlin veered. Losing his momentum, accelerating and closing again, the Merlin climbed above the Dunlin a second time . . . and missed again.

Speed was no ally, so the Dunlin began to use its lighter weight to advantage. It began climbing, spiraling upward in turns that got ever tighter as the birds moved higher. They climbed above the flats and above the shorebirds that had not yet settled. They climbed until both birds were little more than specks, hardly distinguishable from each other.

Then the smaller of the two forms turned and fell. The larger turned and fell faster.

Above the flocks, above the flats, the two birds met and became one. The Dunlin's high, thin scream of protest didn't reach the earth until the bird was dead. Those birds still flying wildly across the flats heard it and began to settle. The cry told them all they needed to know.

Polyborus plancus
CRESTED CARACARA

THE LAND WAS TOUCHED by spring but the sky said "winter."
The bird that was perched on the near side of the pit, like the rain
running in blood-colored rivulets, seemed indifferent. Beyond the
vultures standing on the far side of the depression, cranes were flying. Their bugled cries mingled with the sad sound of geese and the
querulous cries of Killdeer. The bird seemed indifferent to these, too.

And to the multihued cattle ranging across the horizon. And the
rain-colored deer huddled beneath thorny trees. And the mindless
clucking of fowl. And the putrid reek that wafted from the pit.
Indifferent and aloof.

The bird, a Crested Caracara, was a scavenger, a carrion eater, and
today (at least) less concerned with the living than the dead. And like
death, scavengers are an impartial lot, grandly indifferent to everything — except hunger. Even the loftiest of kings are ruled by the
tyranny of their stomachs.

That is why the bird was waiting by the pit, in the rain, on the great coastal plains of Texas. Waiting as the host of Black and Turkey Vultures were waiting — for the tribute that was theirs by default; for the flesh the predator who was their benefactor would deliver. Waiting, as scavengers the world over wait.

Waiting for the kill.

The caracara, an adult, had arrived an hour after sunrise. A large, colorful raptor with a projecting head and planklike wings, it was a powerful bird, a powerful flier — more powerful than the vultures that moved aside at his approach. The suggestion of power was projected by the caracara's eaglelike bill. It was cast in the bird's legs, which were sturdy and long, and in his feet, which were taloned and broad — quite unlike the weak feet of vultures.

When the caracara moved to investigate the contents of the pit, he strode more than walked, and his movements recalled the stalking cadence of the Secretary Birds that haunt the grasslands of East Africa. In fact, the whole scene — the cattle, the cautious ungulates, the vulture hordes gathered around the water-filled hole — recalled the life-and-death pageantry of the Serengeti Plain, and why should it not? In Texas, as in Africa, there are predators who harvest more than they need and scavengers whose job it is to make use of the rest.

It didn't take the caracara long to deduce that the pit was empty. Except for a few picked carcasses floating in a puddle, there was little to tempt a scavenger's palate. He lifted one of the sodden bodies out of the soup, discarded it, then charged several vultures whose presence he suddenly found irritating, claiming their spot.

The vultures flew to the depression's far side and resumed their vigil. It was an old game with them. As Black Vultures are dominant over Turkey Vultures at a kill, caracaras are dominant over both. Being the stronger and more aggressive scavenger, the caracara would have first rights to whatever the predator left, and he was merely asserting his claim and taking his due.

But vultures are endowed with their own wisdom. They know that even the hungriest caracara cannot dominate a kill forever. When the bird's back was turned or his hunger satisfied, their turn would come.

That is why they retreated only a little way. To wait the patient wait of scavengers, who know their business and their place, and

"*The caracara would have first rights to whatever the predator left.*"

The Wind Masters

who know, most of all, that all things come to those who wait. Waiting just as the caracara was waiting.

Waiting for the kill.

For a moment, the caracara considered leaving the pit and foraging elsewhere. But he had already searched the highway for road kills, had left his roost at dawn before the vultures were aloft, finding nothing.

It was a ritual the caracara followed every morning, and on most mornings his forays met with success. The vehicles that sped down Texas roads at night, pickup trucks driven by ranchers and the torqued-up sedans filled with joy-riding teens, killed indiscriminately and cared nothing for the carnage left in their wake.

The caracara might also have hunted. Though carrion figures prominently in the diet of caracaras for at least a portion of the year, the birds are skillful hunters, capable of capturing a variety of prey, including skunks, rabbits, prairie dogs, and assorted rodents; birds as large as ibis and as agile as buntings; reptiles, amphibians, fish, insects, and crayfish. Caracaras come by their hunting skills honestly because, also unlike Black and Turkey vultures, caracaras are *true* diurnal raptors, most closely related to falcons. But where evolution has made specialists out of most of the Falconidae, it has made a generalist and an opportunist out of the caracara.

That is why the bird waited. He knew that the predator who left the remains of its prey in the pit was dependable. He also knew that morning was the predator's period of greatest activity. If the creature held true to form, and it usually did, it would appear within the hour and it would appear bearing prey.

So the caracara waited. Waited as the vultures waited. Waited as Crested Caracaras in the grasslands of Florida and the deserts of Arizona and south into Mexico, Central America, and South America wait. Waited as scavengers the world over wait.

Waiting for the kill.

To amuse himself, the caracara started to preen. With care he ran a pale bill through the mass of body feathers that enfolded the bird like a sleek, black shroud. Lovingly, he fussed with the feathers that covered his neck, breast, and back like a cream-colored cowl and worried the spangled border until he was satisfied.

45

The caracara's crest, which was shaggy and black, billowed in the wind. It might, in keeping with the regal bearing of the bird, be likened to a mane. But in undermining truth the bird's feathered crown more nearly resembled a badly fitted toupee.

Despite the shortcomings of his headpiece, the caracara was a handsome bird, far more attractive than the skull-faced assemblage standing nearby. But the caracara's most arresting characteristic was undoubtedly the bright, bare facial skin extending from his nares to his eyes, and this trait was eminently vulturine. Like the bare heads of vultures, the caracara's unfeathered face allowed him to feed cleanly upon fouled carcasses and amidst the insect larvae associated with them.

The bird's electrifying face served another function — communication. When quiescent, a caracara's face may be yellow, orange, or carmine red. But when excited, agitated, or aroused . . .

A second caracara approached the pit directly and lit, facing the adult. Standing erect, the bird tossed its head until its crest touched its back and gave a hoarse, rattling cry. The adult, its face now flushed to an angry red, mirrored the intruder's overtures and echoed its cry.

The newcomer was a female, although this determination could not have been made by a human observer. Male and female caracaras are virtually identical in size. But the newcomer was also a younger bird, not even a year old. This was very apparent.

Immature caracaras are charcoal brown, not black. Back feathers are buff tipped, giving the birds a spotted appearance. The underparts are heavily streaked. In contrast to the angry red of the adult's face, the younger bird's face was cool blue.

The head-tossing display continued for almost a minute and then stopped. It seemed almost as if some understanding had been reached — the younger bird evidently thought so anyway. But when she advanced toward the pit, the adult dropped his head and charged like a bull for the matador's cape, driving the interloper to the other side of the pit — the side occupied by the vultures.

The younger bird salvaged her pride and secured her place in the pecking order by charging the vultures, who flew a short distance into the field or claimed perches on a nearby fence. The vultures were nonplussed, maybe even amused. They were professional scaven-

The Wind Masters

gers, after all. They knew the game and they knew their place. They could afford to be generous.

The vultures were content to let the youngster buck the line, because patience is a virtue among those creatures who are the ultimate inheritors on earth. So long as there are killers to kill and victims to die, good things will come to those who wait. So the vultures were waiting.

Waiting for the kill.

It was the deer beneath the trees who saw the predator first. They snorted, raised their tails in alarm, and retreated into the trees. Next, it was the Killdeer who raised their warning cries, and then the cattle added their lowing voices to the din.

The vultures heard it, too, the telltale ring of metal on metal that always announced the imminence of the creature who served them. From half a hundred perches they lofted into the air and made for the pit, knowing that the time for patience was done. Some of the younger, less experienced birds flew directly to the pit, vying for position, but the wiser ones held back.

The young caracara did not feel kindly toward the swarm of new arrivals. She charged several vultures, who simply cantered to one side. In the process she lost her more strategic place. The adult, who was in good position, stood his ground and waited. His eyes were fused to the corner of the structure where the predator invariably appeared, and he was not disappointed.

Head down, arms hanging, booted feet sinking deeply into muddy manure, the predator rounded the corner and made for the pit. The feathered host rustled with excitement; several birds were forced into the pit by the push of those behind.

The predator lost his footing at a particularly slimy stretch and went down on one knee, spilling the contents of one of the buckets in his hands — the lifeless forms of half-grown capons, chickens for the market.

"Fornicate this mud and this stupid job and these stupid birds that are too stupid to live long enough to die as they should," he said in very colorful Mexican Spanish. The man was not the owner, merely a hired hand. His job was to tend the coops and to remove whatever poults did not survive the night.

47

He started to retrieve the fallen birds but decided against it. "Let the buzzards get them," he decided. "Why should I do their work, too." He did not like vultures, but he did not particularly hate them either. The caracara he liked because it was bold, an attribute that he elevated in his human mind to "fearless," and because it was, after all, the national bird of his native country, which, even after ten years in the United States, he still missed.

He picked up the buckets once more and made the rim of the garbage pit without further incident. Cowed by the predator's approach, the young caracara and many of the vultures were forced to move off — just as the older birds had foreseen. The adult caracara put a little more distance between himself and the predator but generally held his ground, watching intently as the creature tossed chicken after chicken into the pit.

His buckets empty, the last chicken in his hand, the man made a motion to toss it beside the others but checked his throw. On impulse, he lofted the bird across the pit, toward the waiting caracara. The offering fell short but not so short that the caracara did not jump back in surprise. Once again, his face flushed red.

"*Para ti*," the man said, waiting to see how his gift would be regarded. The caracara studied the chicken and then the man and then the chicken once more, but he did not advance. He was a scavenger. Time was on his side.

The man, who had hoped for better rapport, was disappointed. "They are, after all, no better than the other eaters of carrion," he said tiredly, turning, walking away, oblivious to the flurry of wings behind him. And in a sense, he was right. The birds around the pit were all equals, none better or worse. But a member of a species that gets its dead flesh plastic-wrapped at the market, the same dead flesh that scavengers feed upon (just not as fresh), might think twice before casting aspersions upon the carrion eaters of the planet. The truth is, there was more commonality between the man and the caracara than most humans might comfortably care to contemplate.

A Turkey Vulture and a Black Vulture both made for the chicken closest to the caracara, but the hawk stalked forward, putting both to flight. Bending, grasping the bird with his bill, the caracara claimed

The Wind Masters

his prize and walked away from the pit, taking a position to one side. Planting both feet on the poult, he began tearing off morsels and tossing them down. As he fed, the chicken diminished and the caracara's crop bulged until it resembled a great, orange tennis ball projecting from the hawk's neck.

It took fifteen minutes for the chicken to be consumed and for the bird to conclude that what little remained was not worth his effort. He would leave the rest for whatever carrion eater would have it — and there were many.

Leaning forward, the bird lifted off, climbing little but flapping strongly. Several Killdeer fled at his approach, but unnecessarily. His indifference was almost total now that the waiting was over and he had fed. The predator had done its job well.

There was no need to kill.

EIGHT

Chondrohierax uncinatus
HOOK-BILLED KITE

THE SUN WAS LOW in the western sky — or would have been but for the overcast. The adult Gray Hawk and the adult Hook-billed Kite, both males, had chosen perches beneath the canopy of mesquite that offered some protection from the wind.

The birds were united in their hunger, their aversion to cold temperatures, and their preference for riverine woodlands and tropical forests. They also shared similar ranges — northern Mexico to southern South America. But aside from these broad generalities and a superficial similarity regarding plumage — gray plumage, barred underparts, banded tails — the two raptors had little in common. Their shared proximity, in the riparian and *matorral* woodland of the Rio Grande Valley, the northern limits of their range, was serendipity.

The Gray Hawk, who was a buteo and who looked every inch a bird of prey, turned his dark eyes upon the forest floor, searching for something to distinguish himself — a lizard, a snake, maybe an incautious ground dove — the typical prey of this versatile, tropical raptor. The hunting had not been good. The cold air mass mantling southern Texas had driven cold-blooded animals into seclusion, putting a hungry edge on the hawk's typical raptorial demeanor.

The kite was something else, something of an anomaly. He was slightly larger than the Gray Hawk and less stocky — proportioned like an oversized bowling pin. And though raptorlike enough, the bird had anatomical peculiarities that might charitably be called "interesting," although "ungainly" might serve accuracy better. First there was the bird's bill. It was bulbous and deeply hooked, somewhat eaglelike. But there was something about likening the bird to an eagle that did not inspire confidence, something that shouted "fraud." This suspicion was supported by the bird's legs. They were sturdy but short, giving the bird a sawed-off appearance. As appendages they served to anchor the bird to his perch and offered parrotlike mobility, yet as the defining attribute of a proud bird of prey, the legs seemed poorly cast — too short in the shank to snatch birds out of the air or to pull a rabbit out of the sage (as a "proper" raptor should).

But the characteristic guaranteed to make even a casual observer stop and stare was precisely that — the bird's stare. The eyes of a Hook-billed Kite are startling. Bette Davis wide and billy goat crazy. The pupils seem to swim in a sea of pale green amazement, and the absence of a bony shield above the eye, the anatomical feature that gives most other birds of prey their nefarious demeanor, does nothing but accentuate the Hook-billed Kite's look of dazed amazement.

For all his ungainly attributes, however, the kite was very much a predator; was, in fact, a specialist. And if specialization can be construed to be an evolutionary advance, then the kite was clearly superior to his more versatile, but hungry, cousin, the Gray Hawk. Unlike the Gray Hawk, the kite was surrounded by a surfeit of prey, one that his peculiar specialization allowed him to exploit.

Leaning forward, letting gravity do the work, the bird left his perch and more or less collided with the trunk of a neighboring tree. Bracing with open feet, fending off gravity with forest-honed wings, the kite reached out with his bill, plucking a dime-sized snail from the bark, then flew to a nearby perch. Why this particular snail had won the bird's favor was hard to say. There were, in the immediate

"*For all his ungainly attributes, the kite was very much a predator.*"

The Wind Masters

vicinity, several *thousand* snails and no shortage beyond. They hung on branches and they clung to trees. Set against the gray-green tree bark, the tiny mollusks were as obvious as chalk marks on a blackboard and hardly more mobile.

Terrestrial snails are not just the favored prey of Hook-billed Kites; over all of the bird's range and with few exceptions they constitute the bird's *only* prey, one that this forest raptor is well suited to exploit. Transferring the snail to his left foot, the bird pierced and peeled away the operculum, the dried cap covering the open end of the snail's shell. Then pinning the shell to the perch, he began chipping away the shell, starting at the open base, working down one side toward the tip, opening a channel that exposed the inner whorls.

Chip, chip, chip, chip . . . ten chips and fifteen seconds later it was done. The bird reached down with his namesake bill, extracted the mollusk, and tossed the morsel down whole. As he relaxed his grip, he dropped the shell two feet to the ground upon a pile of empty shells. Evidently, the perch was a favorite.

It might seem odd that a bird of prey, a creature with talons and a flesh-tearing bill, would deign to dine on a creature with the defensive prerogatives of an acorn. But from both a dietary and a utilitarian standpoint, it made perfect sense. The snail had not, after all, required the bird to wait hours, as the Gray Hawk was obliged to do. It did not demand a great expenditure of energy to find or to capture — no long tail chase, no grappling struggle, and virtually no risk of injury.

It certainly required no great effort to reach the animal within the shell. Ten easy snips of a tailored bill. No feathers to pluck. No hide to pierce. No tug of war with tendons and ligaments. Nothing to it.

And if the mollusk hadn't provided more than a mouthful, what of it? There were plenty more where that one came from. And since eating snails is such a specialized art, there were not even any other raptors to come and offer competition. In fact, even among Hook-billed Kites there are birds with different-size bills — one large, one small. These seem to be adaptations allowing birds to exploit different-size snails, thus reducing competition among members of the same species — provided these large- and small-billed birds are the same species (the ornithological jury is still hung on this decision).

In the Rio Grande Valley, only the small-billed variety of Hook-billed Kites is found. And in the park where the male foraged, there was only one other. The bird's mate.

The other kite came in silently. Even the noisy Green Jays, who would have raised hue and cry if a Cooper's Hawk had appeared, ignored her. What concern of theirs was a snail-eating bird? In size and shape the female kite was only slightly larger than the male, but her plumage differed markedly. Cold brown on the back instead of gray. Burnt orange barred across the breast. An orange-colored collar encircled the back of her neck where the male showed white. Hook-billed Kites also have a dark form in which both sexes are similar, and the young of normal-plumaged birds resemble the adult female except that the collar is white and the eye somber brown.

The female bird had not been far away — but in the riparian woodlands flanking the Rio Grande that would be hard. The state park in which the birds were resident was only 525 acres. That and the surrounding woodland were enough to sustain one, maybe two pairs of kites. Since the turn of the century 99 percent of the native riparian woodland along the river has been destroyed for agriculture, development, or recreational purposes. The ten to twenty pairs of Hook-billed Kites found along the Rio Grande between Santa Ana National Wildlife Refuge and Falcon Dam, the products of a range expansion that began in the 1970s, may represent the holding capacity of the land in its present vegetative state — and all the breeding birds found in the United States. This makes the Hook-billed Kite the rarest of North America's nesting raptors.

The female lit in a nearby mesquite, not ignoring but not acknowledging the presence of the male. It was February, after all, and real courtship would not begin until April. She shuffled along the limb, searching for prey. Dropping to a smaller limb that bowed beneath her weight, she leaned forward, turned turtle, and, swinging like a sloth, plucked a snail from the branch. Flying the short distance to another tree, she transferred the snail to her left foot, just as the male had done, and began chipping at her prize.

The pair foraged methodically, mechanically, but this evening, at least, there was also a sense of urgency in their approach to their

The Wind Masters

task. Winter cold spells are not unheard of in the Rio Grande Valley, but the air mass mantling the valley was exceptionally large and exceptionally cold — the coldest in two decades.

The pair fed until sunset, the event marked by an angry red line that defined the horizon and the edge of the cloud layer. Crops full, their energy needs met, the birds flew toward the river and the dense stand of palms that served as their roost.

The Gray Hawk remained until just before dark and then departed — a hungry gray shadow flying swiftly and directly to roost. It wasn't lack of skill that had defeated his efforts to feed. It was the temperatures that had turned his reptilian prey to stone and sent birds to early roosts. It was circumstance and bad luck — the luck of a raptor. Predation is a gamble, and even hawks that cover their bets by spreading their focus across a range of prey sometimes go hungry.

During the night, the mantle of clouds drifted south and the wind died, letting warm air rise and cold air settle. Temperatures that had not reached 40 during the day fell through the 30s, the 20s, into the teens. Cold filled the valley, burning the life out of leaves, and reaching deep into the hearts of hapless plants. It probed beneath the litter layer for cold-blooded things and closed a chilling fist around the shells of snails, turning their bodies to stone and later, after temperatures had moderated, to liquid.

The next morning the Gray Hawk surprised a White-winged Dove whose hunger had mastered her caution and her speed. In the woodlands, upon limbs brittle with cold, the kites were active as well but vexed, confounded by the snails that had lost their succulence and would not surrender their shells. And although they could not know on this the first morning following the great freeze, their time of trial was just beginning.

Specialization, the kites were about to learn, is a two-edged sword. Creatures live by it, and creatures die by it, too.

HOOK-BILLED KITE

NINE

Buteo albicaudatus
WHITE-TAILED HAWK

WIND FED THE FLAMES but contested every foot of the fire's progress. The smoke rising into the clear, blue Texas sky was white . . . except closer to the fire, where it glowed pink, and where the flames tarried over piled debris. Here the smoke roiled ugly and gray.

But mostly it was white. The ranch hands standing near their pickups, monitoring the progress of the fire, noted this with satisfaction. White smoke was evidence of a dry, well-oxygenated burn. This is what it took to keep unwanted growth in check and the coastal prairie in grass.

Two hundred feet above the flames, just ahead of the wall of smoke, a group of hawks was hunting. Poised on obscenely long wings, they peered into the fire and looked for all the world like a host of soot-darkened devils. Not a few creatures fleeing the flames would have regarded them as such.

The hawks, nearly thirty in number, were White-tailed Hawks — a large, buteo species whose North American range is limited to the coastal prairies and adjacent savanna-chaparral rangeland that lie between the upper Texas coast and the Rio Grande Valley. The burn was set in Kenedy County, on the famed King Ranch, right in the heart of White-tailed country.

Most of the birds hovering above the flames were nonbreeding birds hatched the previous season, now in their second calendar year. They had no territory to defend or to define their movements — wouldn't have until the following year, the third calendar year (but second spring) of their lives. Among their ranks were several unattached two-year-old birds whose plumage, give or take a little mottling under the wings, was nearly identical to that of the adult White-tailed perched well ahead of the flames. This bird, a male, was one of fewer than 500 pairs estimated to reside in Texas (or in the United States — a geographic juxtaposition that in the estimates of both White-tailed Hawks and other native Texans comes to the same thing). The fire was on the edge of the adult bird's territory.

From his telephone pole perch, the adult took it all in — the fire, the ranch hands, the younger birds above. Where many raptors are intolerant of other raptors in their territories, White-tailed Hawks are, for the most part, very open-minded about territorial boundaries. Even now, in March, with his mate on eggs in the pair's large, low-placed nest, the bird was nonplused about the influx of young birds. The nest was not close and the adult knew that the intrusion was temporary. Fire is the catalyst for a whole new social order among White-tailed Hawks, and when flames sweep the lands, driving prey from hiding, the birds gather to the feast.

The larger the fire, the more hawks seem drawn to it and the farther they will have come. As many as seventy White-taileds have been seen hunting a burn, but groups numbering less than a dozen birds are more typical. The thirty birds working the fire had gathered over the course of several hours, and many were hangers-on from a previous day's burn. A few had traveled more than eight miles, drawn by the size of the plume and the promise of easy prey. By the look of them it might be concluded that they had walked through the chaparral to get there instead of flown over it. After nine months on the wing, nine months facing down the incessant Gulf Coast winds, the plumage on most of the young birds was in tattered disarray, the feathers severely abraded.

WHITE-TAILED HAWK

"They peered into the fire and looked for all the world like a host of soot-darkened devils."

The Wind Masters

These second-year birds, still bearing the feathers they'd worn upon leaving the nest, were mostly brown — brown above, brown below. Only the barred, gray tails were out of pattern. Each bird sported a prominent white patch on its chest and showed a pale wash bleeding through the dark feathers of its face. Each bird was missing one or two inner primaries, and most were missing tail feathers, too. The gaps in the apposing wings were symmetrical — of course, for balance.

Though opportunely attracted to fire, White-tailed Hawks are allied to fire's great friend, the wind. No bird of prey, not the falcons with their distance-nullifying wings, not the gravity-cheating vultures, not even the frictionless kites, is more at home in moving air or seems better able to turn it to its advantage. Flat, open habitat is the terrain of choice for White-tailed Hawks. But they could not exploit its opportunities were it not for the wind, which is to the White-tailed Hawk what a perch is to other birds.

The flight of the White-tailed Hawk is buoyant, Swainson's Hawk–like, but more acrobatic. The birds can glide like a heavy-bodied Ferruginous Hawk, hover like a Rough-legged Hawk, and swoop like a Mississippi Kite — but such comparisons are for the benefit of human observers who may never have seen a White-tailed Hawk. In truth, the bird's execution of these aerial maneuvers makes the efforts of most other raptors seem stillborn.

White-taileds do not merely glide. They can edge through air with the ease of a whisper.

Not only can White-taileds hover, they can do so by teasing the air with their wing tips instead of resorting to an arching, energy-sapping flap.

And as for swooping . . .

Coursing above the flames, searching the unblackened terrain ahead and scorched earth behind, one tattered juvenile-plumaged bird suddenly drew in its wings, turned, and *plunged* . . . headfirst; straight down. The movement was so sudden, so neatly executed, that there seemed no movement at all. One minute the bird was gliding forward, the next it was arrowing down. Nothing to it (so long as you are a White-tailed Hawk).

Maneuvering half-opened wings into a controlled, corkscrewing

59

glide, the bird turned back into the wind and stalled. Opening his wings benediction-wide, raising them like the arms of an Olympic diver poised on a high board, the bird held this pose for several seconds, then floating more than falling, he eased his extended talons through ninety feet of air, burying them in a gopher snake that had escaped the flames.

Lifting the animal beyond the reach of the heat, evading the harpylike efforts of two other young White-taileds, the successful hunter flew to the springy top of mesquite and started to feed. White-taileds will perch-hunt when there is no wind — in fact they quickly resort to this tactic where trees lie in the path of a fire. But it is the absence of trees that gives the birds the strategic advantage they need both to secure prey and to out-compete other open-country raptors. Red-tailed Hawks, Ferruginous Hawks, and Northern Harriers are their principal competitors.

Abruptly the adult took wing. Flying toward the advancing edge of the blaze, opening his wings to embrace the burn, he drew aloft behind a curtain of pink and gray smoke. In plumage, even in shape, he seemed vastly unlike the other birds. In fact, if someone were to suggest that the dark birds and the adult were entirely different species, it could hardly be disbelieved.

Almost as long-limbed as the younger birds, the adult displayed wings that were broader in appearance and more uniformly tapered — less pinched at the body and wider at the hand. In a soar, in comparative fact, the bird seemed to *be* all wing. The fully fanned tail, fused to the trailing edge of the wing, hardly seemed a separate appendage at all.

In shape, he recalled the Bateleur, an African snake eagle whose habitat, perhaps not coincidentally, is also dry arid plains. In plumage he was most like an adult, light-form Short-tailed Hawk — with one qualifying difference. He was handsomer.

Bursting through the smoke, set against the backdrop of the billowing plume, the underparts gleamed like a fair-weather cloud, white on white. The brilliance contrasted with, and was set off by, the charcoal trim of the bird's wing and a broad, dark band on the tail. The plumage was fresh and trim; courtship-fine. Not a feather was out of place.

The Wind Masters

The bird's head and back were pearl gray, the shoulders russet and blush. The sides of the breast and the underwings were finely and warmly vermiculated, and the hawk's expression, though none could see it, was patient, almost serene (as befitting a creature in total harmony with his environment).

In a word, the bird was beautiful, very possibly the most beautiful raptor in North America. But to focus on appearance alone is to sell the bird short. If subadult White-tailed Hawks exhibit finesse in the air, then there is only one word to describe the poise and movement of an adult White-tailed Hawk.

Masterly.

The bird stopped rising, turned into the wind, and stopped. He did not hover. He did not move at all. He simply stopped — rested his weight on the wind, suspending the laws of physics. The bird remained there like a held breath, like suspended judgment, like Icarus before the fall — searching the earth ahead of the fire for the next flame-flushed creature. It seemed that the laws of physics were destined for a rewrite when the bird plunged — a loose-winged, arching spiral that fused with the wind.

The bird passed through the cloud, becoming part of it. Sometimes the veil of smoke parted, offering glimpses to mortal eyes. What could be seen was a plummeting form that hesitated between fluid and vapor and only occasionally appeared in earthly form. He reached the ground before it seemed physically possible and was climbing with prey before heaven and earth could work out the terms of exchange.

Even the young birds who left the adult unmolested could not say how the bird had done it. Only the vole clutched in the hawk's talons confirmed that it had happened at all.

Climbing until the heat of the flames was insignificant and the wind and his weight reached parity, the adult crooked his wings and set a course to intercept the nest and the mate who would be watching the smoke, and waiting.

Buteo jamaicensis
RED-TAILED HAWK

THE VALLEY WAS STILL IN SHADOW when the bird left the perch where she had spent the night and headed for the far side. From his roost perch close to the old nest tree the male watched her go.

She flew steadily because there was no lift to help her — over the pond, over the rutted lane, and across the overgrown field that filled the Appalachian valley floor. Only when she approached the tulip poplar that dominated the far woodland edge did she set her wings and glide. Her outstretched talons and the cresting light of a winter sun touched the favored limb simultaneously. Once again, her timing was perfect.

The male watched this, too. And how she paused. How she turned. How she faced him with the full light of morning upon her — a pale-breasted bird with a blush-colored tail and a dark band of dark feathers etched across pale white underparts.

He watched as she fluffed her feathers to buffer against the chill.

He watched how she raised one taloned foot and buried it in her breast; how she ran her bill down the length of an unruly primary and preened along the sides of her breast, and finally, her morning ritual completed, how she stood perfectly still — a large, adult female Red-tailed Hawk in the prime of her life, his mate.

With or without the sun she was beautiful.

Very soon, the male would fly across the valley to join her, and he was eager to join her. This had not been the case in August, when their brood from the previous season had dispersed, ending that season's domestic duties. Since that time, and until a month into the new year, both birds had behaved much like a human couple whose attraction for each other has been eroded by time and supplanted by routine. They shared the same valley. They were aware of each other's existence; drew comfort and support from it. But they led separate lives and were content to maintain their distance.

Over the southern portions of the species' range, a range that encompasses almost all of North America south of the tree line, this is how Red-tailed Hawks spend the winter. Northern birds, those heralding from Alaska, Canada, the northern Great Plains, and portions of New England, generally retreat from breeding territories when the first snow flies, and though they have mated for life, the diffidence that falls between paired birds is sharpened and enforced by distance.

But even before winter's cover of snow has turned to slush, the geographically estranged pairs that are winter fixtures along highways and across pastures have started north. Their paths carry them over the territories of southern kin whose harsh, whistled cries rise to meet them.

The male bird was *very* eager to join his mate, had been growing ever more aware of her, in fact, since January — when increasing amounts of sunlight reaching the earth had stimulated the bird's pituitary glands, triggering hormonal changes that cast the pair's staid social arrangement in a different light.

He wasn't content to watch from afar, merely to roost in close proximity. He wanted to be near her. He wanted to fly with her, over the several square miles of woods and fields that was their home range. He wanted to perch with her, call to her, touch her, and be touched by her, *but* . . .

But he was reluctant, too. First, because she was much larger than he was, large enough to inflict injury if she chose. Second, she was not as receptive to the rekindling of their bond as he was, not yet.

And third, because even though they had courted and mated before, even though they were a pair of several seasons, in a very real way their courtship was for the first time — the first time again. The hormonal changes that were occurring in both birds were roughly equivalent to going through puberty. All that had passed between them was past. What lay before them was the ritual of probing, testing, and seeking assurance that is called courtship.

Think about it. Think back upon your own human experience in that long, nightmarish period called adolescence. Nobody ever forgets the trials of that eventful period of biological development. And nobody I have ever questioned wanted to repeat it.

Do you recall how once you sat in that totally boring class treating that deathly dull subject and stared at the back of the head of the classmate one desk ahead — the same classmate you'd known since kindergarten? Do you remember how one day that classmate turned, smiled, and asked to borrow a pen, and how you wanted to answer but your tongue was suddenly about two sizes too big for the inside of your mouth and all you could do was rummage in your desk for a pen and *stare* at those incredible, incredible eyes?

Funny you'd never noticed them before.

Do you remember how you used to keep track of eyes-so-bright in study hall? How you went out of your way to learn what his or her class schedule was and where he or she sat? Do you remember how you used to maneuver to be near — in the cafeteria during lunch . . . in the hall between classes . . . on the bus going home — maneuver to be *close*.

You were afraid to go up and find out whether that bright-eyed focus of your life liked you as much as you liked that person. Because what if he or she said no? You'd be crushed. Or worse. What if the answer was yes? Then what? Then . . .

Sitting with each other on the bus and sharing chewing gum.

The Wind Masters

Exchanging rings and secrets and notes folded so many times that they formed paper cubes. Hands found hands eventually and maybe lips, too. But the next thing you knew, those incredible, incredible eyes were falling on someone else. Your world was in free fall. Your self-image crashed with it. And you could have just died.

Now, can you imagine being condemned to go through this paralyzing torture every spring for the rest of your life?

The full sun vaulted the horizon, flooding the valley with light. The female fairly glowed with it — the only patch of brightness against a winter landscape; the only thing the male Red-tailed had eyes for. Abruptly he left his perch and started across the valley.

Beginning in early January the male had begun altering his morning pattern. Though the birds had separate hunting perches and different patterns, he began following the female, claiming a perch in a nearby tree. The female had accepted this.

Several days later he had started taking perches in the same tree. The female had accepted this, too.

Then, on a morning boasting unseasonably high temperatures, the male Red-tailed had pushed the season and the female's tolerance past threshold. He had alighted upon her perch. *Her* perch! Exacerbating the problem, he had chosen to test her compliance while she clutched newly captured prey, a vole, a woodland *Microtus* — the prey célèbre of Red-tailed Hawks everywhere.

The female would have none of it. She reared up, towering above the one who had been her mate, and expanded the feathers of her chest, exaggerating her size. Her eyes bore into his and she took a step toward him, crowding the perch.

The male took the hint. He left. Claiming another perch in a nearby tree, lower down than the one upon which she sat — a gesture that was both submissive and apologetic.

Who is to say whether the rebuff was still on the male's mind as he approached the tree again? Certainly he had waited several days before putting her feelings to another test, though he had been anything but low-profile. During the day his perches had been conspicuous, chosen with care, rarely out of sight of his intended. When soaring conditions were at their peak, he had climbed high above their territory, advertising his existence and his claim to all Red-tailed Hawks in the valley — but particularly to one.

Maybe he was building up her confidence, trying to assure his once and future mate that his interest was sincere and that there was no threat. Or perhaps the confidence he was trying to bolster was his own.

The reversed sexual dimorphism evidenced in birds of prey, the tendency for females to outsize and outweigh males of the same species, has puzzled ornithologists for many years, and many explanations have been offered to account for it. It has been suggested, for example, that males and females differ in size in order to utilize a greater range of prey, thus reducing competition between them. Females do, in fact, capture larger prey on the average than males.

It has also been shown that the species in which dimorphism is most pronounced are those that are the venatically dynamic members of the raptor tribe — harriers, Harris' Hawks, and especially bird-eating raptors such as the accipiters and falcons. Conversely, the species that exhibit little size difference between males and females are species such as the Snail Kite, Black Hawk, and American Kestrel that consume a great deal of easy-to-exploit invertebrate prey.

The evolutionary benefits of dimorphism have also been linked to raptor nesting success. It has been suggested that evolution favors larger females because superior body size better meets the energetic costs of producing eggs or the insulating needs inherent in incubation. Others have argued that females must necessarily be larger in order to defend their young from the infanticidal inclinations of males.

Others, conversely, have contended that small size offers selective advantage to males who must out-compete rival males in aerial combat or that small size helps males to provide food for a mate and young insofar as small prey items are generally more numerous than larger prey.

All of these arguments have their strengths. Some have been shown to have weaknesses. Most may be, to some degree, correct because evolution does not necessarily turn upon a single determining factor.

But there is another possibility — one that does not focus upon prey, or nesting adaptations, or the relationship of birds to their young. It focuses upon the relationship of paired birds to each other, or more specifically the relationship before the birds become paired.

The Wind Masters

The difference in size between males and females may be essential to establishing the pair bond between two independent and predatory entities. It may be that the female needs the security inherent in her size to feel comfortable and confident before the advance of the male; that the male's dangerous potential must be tempered by a healthy respect. Only from this secure footing can the barriers that lie between two independent predatory creatures begin to go down. Just as in humans, the foundation of a couple's pair bond is trust and respect.

The male did not fly directly to the perch. He circled several times, biding his time, taking the edge off his advance. The female who had watched his approach continued to preen. It wasn't much of an affirmation, but it wasn't a direct rebuff either.

His decision made, the bird approached the limb and lit a wing's length away. He was careful not to raise his head, and if anything the bird seemed frozen in a crouch.

For a moment it seemed as though he had miscalculated again. The female drew herself up, pinning the intruder with eyes dilated with anger. But she made no greater threat, and slowly, gradually, when the male made no other advance, as he held his posture and his place, the female relaxed.

The male remained motionless until the female began preening again, then slowly righted himself, still facing the other way. When he was confident that he had been accepted, he turned, slowly, until both birds were looking out over their valley. He did not move closer. She did not move away.

For now, it was enough. If the day became warm, as it promised to, he would get up and soar over the place where the old nest stood and she might join him. If not today, then tomorrow or the next day.

These soaring flights, these prenuptial flights, would increase in frequency and ardor as February slipped away — she leading, he following, two broad-winged birds with chestnut-colored tails drawing ever-tightening circles across the sky.

Sometimes they would touch wing tips, because they wanted to be close. Sometimes the male would touch her lightly upon the back and she would turn in flight and meet his talons with her own.

Morning by morning the distance separating them on the limb

67

"*The distance separating them on the limb would disappear. One day their shoulders would touch.*"

The Wind Masters

would disappear. One day their shoulders would touch. Their bills, too, and in time the curved bills so adept at rending prey would come to preen the other — and if you care to think of this as a caress, there is no one to say you are wrong.

But these elements of courtship would unfold in the days ahead. For now, on this morning in February, the birds were content to just sit. On the same limb. Overlooking the valley that was theirs by another marriage and would soon be theirs again.

ELEVEN

Accipiter striatus
SHARP-SHINNED HAWK

THE REVEREND SAMUELS and the subadult female Sharp-shinned Hawk were both up early and both absorbed in their respective tasks. The Reverend, seated in the study of his suburban Washington, D.C., home, needed an Easter sermon to nourish the souls of his congregation. The hawk, studying the birds coming to the minister's feeders, needed breakfast.

It did not disturb the Blue Jay–sized raptor that the backdrop behind the feeders was a latticework of trees and shrubs. Sharp-shinned Hawks are forest hunters. They can pass through branches the way a shuttlecock eases through a loom.

It didn't concern the hawk that dawn, on this morning in early April, appeared to be breaking simultaneously in the east and the west. Metaphysical anomalies fall within the fief of ministers, not accipiters.

What did trouble the bird was the forty feet of open space that separated the feeders and the last bit of cover the hawk could use to conceal her attack. She knew it was too much. Before she would be able to bring her talons to bear, the birds would see her and head for cover. She knew this as surely as she knew the limits of her abilities and theirs; knew it from experience.

What the hawk needed to close the gap was an advantage — something that would gain her a few more feet before she was discovered; something that would tip the fine balance between predator and prey her way. That is why the bird had left her roost and taken a perch at first light — slipped beneath the branches of the old apple tree, lighting as silently as a feather in free fall.

Brown cloaked, streak breasted, vertically perched, she became part of the tree itself. Only her head, which moved in measured turns, and the orange-flecked yellow eyes were alive. These she used to search the near branches and the yard beyond for opportunity or danger.

Relaxing slightly, she continued her vigilance and completed a bit of morning toilette — ran a bill down the length of several primaries, harassed the feather lice about her head with several licks of a taloned toe — roused vigorously, then settled down to wait. The cloud of feather dust her efforts raised hung in the air like a pale shroud.

She was a hundred feet from the minister's deck and sixty feet from the overgrown hedge that would conceal her during the first stage of her attack. The several bird feeders, her targets, were hung about the deck.

As she watched, the first furtive birds began to arrive. Shadow chickadees and shadow titmice that snatched sunflower seeds and fled. Cardinal, junco, and White-throated Sparrow silhouettes that foraged hungrily for the seed scattered beneath the feeders.

These were the elements of the minister's "other flock," as he called them. Though not an avid birdwatcher, the minister nevertheless enjoyed feeding birds. They brought color to the winter landscape and more than once had served as a source of inspiration for his sermons.

Through contemplation of his feathered minions, the minister had divined such Christian concepts as charity (his) and faith in the Provider (theirs). To visitors and houseguests who marveled at the number and diversity of birds, he was fond of pointing out that God, in his wisdom, had seen fit to craft but two creatures in the Universe with feathered wings — angels and birds. The birds flocking to his feeders served as a constant reminder of the handiwork of the Creator.

In the minister's mind, the similarity between birds and angels was even more acute because just as there were "good angels" and "bad angels" there were good birds and bad ones. The former consisted of the members of his flock. The latter were the rapacious hawks, the Sharp-shinned and Cooper's Hawks that swooped down like demons, plucking the innocent souls from his feeders and bearing them away to torment.

The Reverend Samuels was not the first person to feel protective about songbirds or to think ill of the accipiters. In the previous century, the Sharp-shinned Hawk had been widely regarded as a "bloodthirsty little pirate" (to use the words of ornithologist Warren Eaton). Even during the first half of the twentieth century, there were bounties on the heads of accipiters, and after they were lifted, the protection legislated for other birds of prey was withheld.

Only in the last fifty years has the *un*popular regard for this agile, bird-catching hawk been tempered by an understanding of the important role that *all* predators play in the environment. Even so, most people still find it difficult to accept, without protest, the fate of some favorite feeder regular pinned to the earth by talons. At such times it is difficult to recall that predator and prey are the enjoined halves of the same coin and that each helps maintain the health and stability of the other's population.

Difficult to accept that the inhibitions that prevent members of a species from harming other members of the same species (including ours) have no standing once the boundary *between* species is crossed.

Difficult for people to accept their culpability in another creature's demise. Because, in many ways, a bird feeding station is what our legal system would define as an "attractive nuisance." It prompts birds that would otherwise be foraging widely to concentrate in a small area. Such an unnatural situation is a natural invitation to predators. It was, in fact, precisely this that had attracted the Sharp-shinned Hawk to the minister's feeders in the first place.

She was a second-year bird, heading north for the first time, and although she still carried the plumage of an immature Sharp-shinned Hawk, she was capable of breeding. Home for her was the boreal forest region north of Ottawa, Canada — though the breeding range of Sharp-shinned Hawks covers most of Canada and much of the northern United States and extends south along the Appalachian Mountains in the East and the Rocky Mountains in the West. Winter quarters for the bird were the bird-rich bayberry thickets of coastal North Carolina — although she may have chosen to voyage even farther south, as far as Florida, Cuba, or Central America.

The bird's passage journey the previous September and October had been largely coastal and had antedated the migration of adults by a week. The return flight she was making now would be inland and more direct. This is how she came to be in Fairfax County, Virginia, and how she had discovered and marked the surfeit of birds around the minister's house.

Sharp-shinned Hawks are exquisite predators, and the anatomical refinements that make them so formidable in the woodland arena are the same that distinguish them from falcons, buteos, and other birds of prey. These traits include:

Short, broad wings designed for rapid acceleration across small, open areas and through tight, wooded confines alike.

A long, rudderlike tail that gives a pursuing Sharp-shinned Hawk the maneuverability of a dancer and the bonding properties of a shadow.

Long-toed, lance-tipped feet that can plant death within or squeeze the life out of prey.

Legs with the reflexive response time of a mirror and a reach that is boardinghouse bred.

But a measure of her prowess was also linked to experience. Hawks are not born knowing how to hunt — no more than a carpenter is born knowing how to build cabinets. Hawks are born only with the capacity and the urge to capture prey. The skills must be acquired.

Since relinquishing the care of her parents the previous August, the Sharp-shinned had been acquiring those skills. Every failed capture had been a hard lesson learned. Every success had brought refinement to her technique and taught her the many tricks of her trade.

When the bird had discovered the minister's feeders the previous

73

evening, she had brought her experience and skills to bear; had used the house to shield her attack — hoping that she could turn the corner and surprise the birds at the feeders before being seen. It was a technique she had developed and used effectively in the coastal forests where thickets are plentiful and sight lines are short.

But the minister's yard was more open than the barrier-island cover she was used to, and the birds at the minister's feeders were savvier than the swarms of Yellow-rumped Warblers that wintered along the coast. She was not the first Sharp-shinned Hawk the minister's flock had seen and not the first to employ such a tactic.

While she had been learning the business of being a predator, the birds at the feeders had been learning the art of being survivors, and they had learned it on site. The Sharp-shinned had not even rounded the corner before a perched Blue Jay brayed an alarm. By the time the Sharp-shinned made her cut, the birds were seeking shelter in the trees and bushes that flanked the porch.

She made an effort to grab a House Finch whose escape route carried it precariously close. But the finch was healthy and quicker. After an initial, failed grab followed by a short chase, the hawk broke off her pursuit. Sharp-shinneds are bushwhackers and sprinters, not long-distance pursuit artists.

She perched on the railing of the deck, studying the sparrows huddled in the hedges and the several bolder chickadees who soon returned to the more distant feeders — but she did not mount a second attempt. What was the point? The element of surprise was gone.

She did make a half-hearted charge toward a cardinal whose scolding *pique* was irritating and whose approach was borderline reckless — but the cardinal sought shelter, avoiding her easily.

She stayed long enough to size up the situation and conclude that lingering held no promise, and then she left, to hunt elsewhere. Now she was back.

More and more birds were converging by the minute. They vied for perches and fought for space beneath the feeders. They filled the air with the quarrelsome notes of hungry birds and spent the balance of their energy searching for food. If Sharp-shinneds could smile, she would have smiled because, as she had come to learn, hungry birds are incautious birds. It gave her an advantage and a part of her said, "*NOW.* Do it *NOW.*" But she waited.

The half light of morning was another advantage, as was her concealment, as was her ability to see her target in advance — to search among the birds for weak or injured individuals, birds whose infirmities both hinder and distinguish them when a flock explodes into motion. And among the birds beneath the feeders there were two birds that seemed less able than the rest — a sparrow and a junco. Once again a part of her said, "*NOW.*" But once again she waited.

Her greatest advantage was the hedge that would shield her for more than half of her approach. Her greatest disadvantage was the forty feet that lay beyond. The trick would be clearing the hedge without displaying her silhouette against the morning sky. If she could do this, her chances of reaching the feeders successfully were good.

And tucked away in the young hawk's manual of hunting skills was a favorite trick that could give her the momentary advantage she needed. It wasn't foolproof, but the tactic was one that had worked often enough for her to have developed it into a sort of style. That is why she had arrived so early. That is why she waited.

Suddenly, an alarm note sounded and the porch was nearly obliterated by blurred wings. The concealed Sharp-shinned focused her attention upon the short-lived struggle ensuing beneath the rightmost feeder — the one closest to the corner of the house. Another Sharp-shinned was standing atop a junco and looking in the direction of the apple tree, looking at the younger bird.

The successful bird was smaller, a male, orange breasted and blue backed, an adult. He had used the round-the-corner trick the younger bird had tried the previous evening, but with greater skill and strategic advantage. He had been working the feeders all winter, and in the body language of Sharp-shinned Hawks what he was saying to the interloper was: "This is mine." Despite her anger, and her superior size, the young Sharp-shinned was not about to contest the other bird's prize. She was a patient hunter. She would wait. The minister, however, was more disingenuous. When the bulky silhouette suddenly appeared beyond the feeders, the adult fled — taking his break-

SHARP-SHINNED HAWK

fast with him. The minister, who was intent upon the morning sky, never saw him.

As much as the birds, John Samuels loved to watch the sunrise from the comfort of his porch. The house was old, the porch new — light, airy, glass enclosed — almost more outside than in. In the summer, he sometimes wrote his sermons there.

"Sermons!" he thought uncharitably, watching how the light was gathering in the eastern sky, wondering absently why there were no birds at the feeders on such a cold morning. He had considered and discarded several of the standard Easter themes; had even roughed out one that focused upon the enjoined natures of God and man. How both were irrevocably bound and how Christ's death and victory over death was the covenant.

"He died so we would live," the minister half said to the dawn. He meant it and he believed it. But what troubled the minister about the death and salvation theme was that *everything*, even religion, seemed to be getting more secular. It seemed that everything he'd been preaching from the pulpit had a people theme, not a God theme, and although he believed that religion was supposed to interpret God to man, what he really wanted to do this Easter was raise man to God. What he really wanted to prepare was a sermon that did some good old-fashioned honor and praise to the Almighty.

He checked the level of seed in the feeders, and he saw that it was good. He looked at the eastern sky and watched the clear, cold brightness sweep the heavens clean.

"Every day a miracle," he thought.

He lingered until the sun began cresting the horizon — a strong sun, fueled by spring, so bright it made his eyes squint.

"Every day a miracle," he thought again before turning back toward the den. "And a miracle is what it's going to take for me to get a sermon down on paper," he added as an afterthought. The birds were returning to the feeders en masse before he'd crossed the room.

The hawk saw the minister disappear and saw the sunlight sweep across the yard. Accipiter-quick she turned her head, meeting the sun with a yellow eye, then turned back toward the feeders, taking final note of the distribution of birds. Then leaning forward, she left the perch.

76

Gravity and three quick wing beats brought her down to grass-top level, fully concealed from the birds at the feeders. She leveled off, increased her wing beats to five per second, and started climbing again.

It took less than a second to halve the distance to the hedge and she was traveling at twenty-five miles an hour. Half a second later she was three feet from the nearest branches, a foot below the untrimmed top, and she and her shadow seemed destined to collide.

But hedge hopping is old hat for accipiters, and their reflexes rival a shadow's capacity to mimic. The bird raised her tail, set her wing, and vaulted the hedge in a silent, stealthy, hedge-hugging bound. Her silhouette hung in the air for less time than a sparrow's heartbeat — and though many eyes were watching, none could pierce the halo of morning sunlight that surrounded and shielded the hawk.

It was a favorite trick, hunting with the sun to her back — one that she had discovered and refined in the coastal thickets of the Carolinas. And this time, again, it had given her the concealment she needed.

She was ten feet past the hedge, searching among the feeding birds for a target, when a mockingbird saw the intruder and rasped an alarm. The need for stealth gone, she brought her wings into play for the final dash, accelerating rapidly.

She was twenty feet from the nearest bird before the flock exploded into motion. Most of those who were healthy and alert lifted off and scattered. Several birds who for one reason or another had not pinpointed the approaching hawk had reacted by immobilizing their forms. Among them was a White-throated Sparrow. The hawk fastened her orange and yellow eyes on it.

She was twelve feet from the sparrow before it seemed to come to grips with its peril. Ten feet before it pushed off from the patio, and six before the bird could wheel about and get its retreat under way. The sparrow was flying directly away, making for a sun-lacquered latticework of trees and shrubs that the Sharp-shinned knew she could never reach in time. She knew this from experience.

Four feet behind the sparrow the Sharp-shinned ceased flapping. At three feet the hawk began swinging her feet forward to add the force of their strike to the force of her weight and speed. They were

"*The sunlight streaming through the dust-stenciled form made it glow like a halo.*"

The Wind Masters

almost fully extended just before impact, raised to lie on a horizontal plane that carried the full weight of the bird. Her aim and her timing were perfect.

As the sparrow and the talons met, the hawk opened her wings, fanned her tail, and angled her body upward. Her eyes, her feet, and her prey became linear. It is unlikely that she even noticed her reflection in the glass. It is certain that she had no time to avoid it.

The Reverend Samuels heard a loud bang and returned to the porch to investigate. He entered the room, raised a hand to shield his eyes from the sun, and saw, silhouetted on the sliding glass door, the perfect outline of an angel. The sunlight streaming through the dust-stenciled form made it glow like a halo. Beneath the image was a fallen sparrow.

"Not a sparrow shall fall to the ground that he does not see it," the minister whispered, paraphrasing the words of Matthew 10:28–31. Though saddened by the death of the bird, he was very suddenly very moved by the omniscience of his God (as well as being bewildered by the outline of the angel!). But, to his credit, the Reverend Samuels was not so bewildered that he didn't realize that his prayers had been answered. John Samuels had his sermon.

"God works in strange ways," he said, and he was right, though he did not appreciate the extent of this truth. The bird lying at his feet, the bird whose weakness had made it the target of the hawk, was infected with salmonella — a fatal disease, transmitted bird-to-bird through the droppings. It is one of the hazards birds face when they concentrate as they do at feeders. The bird's removal would prevent the transmission of its disease to the others. It had died; they would live.

In the apple tree, the hawk that was the tool of the Lord's will was trying to make the leg that was broken work as it should. In time, the hawk could learn to function with just one, but before that could happen the hemorrhaging that was already beginning to squeeze the life out of her brain would have to stop. That would take another miracle.

79

DAS

TWELVE

Buteo lineatus
RED-SHOULDERED HAWK

THE STARTER WAS BROKEN, so he jumped the tractor by spinning the flywheel. She caught on the third try, faltered, then fell into a chugging rhythm that brought a rare smile to the face of the forty-six-year-old New Hampshire native.

"Never knew a time you couldn't spark, old girl," he said aloud, because he was alone, completely alone, and because he always spoke to his tractor when they worked. He never actually addressed the tractor as "Anne," the name of his deceased wife. He merely thought of the tractor by this name. The projection had come so gradually that it was familiar before it was conscious — and it might have been that he was not conscious of it even now.

They had bought the tractor instead of the sport sedan she coveted, surrendering in Yankee fashion to practical need. Her diagnosis had coincided with the first payment; her death with the last. Their final journey together on this planet, to the crest of the hill overlooking the farm where her ashes were spread, was aided by the old John Deere.

Checking the hitch, making sure his lunch was in the toolbox, he climbed into the seat, eased the tractor into gear, and pulled out of the shed. It was a good day for plowing.

The lane was mostly dry in the center, muddy only where the April sunshine had been defeated by the pines. A car might have had trouble navigating the worst spots, especially one driven by someone who didn't know how to drive in mud. But it was April now and mud time was drawing to an end. On the hillside sugar maples were blushing with spring, and on both sides of the road, furry-stemmed flowers were thrusting sun-colored heads through a shroud of last year's leaves.

"Coltsfoot's up," he said aloud, so he could hear the thought over the sound of the tractor. "Be phoebes any day," he prophesied as they crossed the bridge over the stream, and, though the tractor moved on, the man's mind remained on the bridge — recalling a day, twenty years ago, and a picnic and a promise nullified by fate, but never broken.

It had been a day like this one when the young Navy pilot, just back from a bad war, and a young woman, just back from college, had eaten store-bought sandwiches and sipped not-so-good wine, and talked of experiences, and old high school friends, and ambitious (but not inflexible) plans for the future — talked, in fact, all around the thing they wanted to talk about most.

It had started well, but wandered badly, and then, with lunch over, and the words still missing and the afternoon grown awkward, the impasse was suddenly broken by the cries of a Red-shouldered Hawk. On frenzied wings, the bird circled above the pair who were not yet a couple, climbing until his outline lost definition and almost disappeared against the sky.

Pausing, plummeting, the bird hurtled earthward, his wings folded back like the wings of the F-4 Phantom the man had piloted in combat. Then pulling up, just above the trees the bird rocketed skyward, climbing now like a kite caught in a wind gust, his crimson shoulder patches blazing. Again and again he climbed and fell, climbed and fell, one small miscalculation away from injury or death. The valley rang and echoed with the cries of a Red-shouldered Hawk in the spring and the sassy mimicry of Blue Jays who taunted him (but did so from the cover of the trees).

"Bet you aren't as good a flier as that," she said without turning.

"Better," he said, knowing that it was not true.

The bird continued his aerial display for ten minutes, taunting Earth's gravity; challenging the sky. And then the female left the hillside, seduced the young ace from the sky, and, lighting upon an open limb in front of the two picnickers, they mated.

"Don't bet I'm not as good as that?" he teased, feeling some of the wine; feeling some of the old fighter pilot's cockiness. She did not respond and he was suddenly sorry. He turned to apologize, only to discover a wry smile playing across her face and a look that was both piercing and probing. She didn't respond to the challenge so much as match it.

Abruptly the man slipped the tractor into neutral, stood straddling the seat, and, raising his hand against the sun, scanned the hillside, searching for and finding the old beech tree that the pair of Red-shouldered Hawks favored with their nest. The birds had used several different trees over the years, including two different red oaks, an ash, even a white pine (although the birds clearly preferred deciduous trees). The beech was their favorite, and it was sometimes used two or three years in succession.

He could not see the modestly sized stick nest, set close to the trunk, cradled by the first jutting limbs beneath the canopy, nor could he see any sign of the birds, and this troubled him. Their courtship should have been well along by this time, and courtship in Red-shouldereds is protracted, four or five weeks. He didn't think he'd missed it.

He'd seen one adult bird but he'd yet to see two. Disappointed, he eased back into the seat and put the tractor into gear. Five minutes later he was in the field, turning the stubble of another season beneath his blades.

From its perch, near the old nest, the adult Red-shouldered saw the tractor but paid it little mind. For seven years he'd watched the ritual turning of the earth and it meant nothing to the bird. Instead the bird watched the sky, searching for the mate who had not appeared. As the man had surmised, she was late.

Over most of the Red-shouldered's eastern range, which roughly encompasses the reaches of the eastern deciduous forests, and throughout the bird's disjunct California range, Red-shouldered Hawks are permanent residents. Only in Canada and the northern states do some or all Red-shouldereds migrate. They leave their nesting territories in October and November; return in March and April.

What the bird could not know but had begun to suspect was that the female, his mate of four seasons, was not going to return, ever. He had been on territory for a week. Their respective arrivals had never been separated by more than three days. If she came, that was well. But if another came in her stead, that, now, was acceptable.

In nature, in Red-shouldered Hawks and many other species of raptor, territory is key, the birds that occupy it interchangeable parts. While the man believed that the Red-shouldered Hawks that occupied the farm were always the same birds, the truth was that two different males and three different females had nested there since the day at the bridge. Going back even further, the forested hillside could claim unbroken residency by Red-shouldered Hawks for eighty years. Once New England's forests began recovering from the logging of the last century, Red-shouldered Hawks returned and thrived. The woods above the farm offered all a pair of Red-shouldered Hawks could ask: a large, unbroken woodland dominated by maturing hardwoods; nearby fields with a surfeit of edge and a forested stream; the favored hunting grounds of the Red-shouldered Hawk.

Over much of the South, Red-shouldereds are associated with hardwood swamps, though in Florida and Texas they also favor pine woodlands. In the western reaches of their range, where the eastern forests surrender to prairie, Red-shouldereds utilize riparian habitat just as they do in California.

Wherever nesting Red-shouldereds are found, occupied nests are decorated with a sprig of greenery. It might be the leaves of an early-blooming flower such as the violet. It might be a branch cut from an evergreen. Whatever its source, its message is clear. It says to prospecting Red-shouldered Hawks, "This nest is occupied."

The nest in the beech on the New Hampshire hillside bore no such sign, and this, in the language of Red-shouldered Hawks, is as good as saying "Vacancy." There is another way Red-shouldereds communicate this message, and abruptly the unmated bird left his perch, weaving a path through branches and trees in the manner of a consummate forest buteo.

He cleared the canopy, gained the open air above a granite outcropping, and fanning wings and tail caught a good, strong spring thermal that carried him aloft.

In flight an adult Red-shouldered is a compact bird — a plank-winged, broad-tailed raptor whose lines are trim, dress-uniform fine. Spring sunlight glowed through the narrow white bands on the tail and melted through the tips of the wing, branding each with a pale crescent. Only when the bird banked could earthbound creatures catch a glimpse of the russet upper wing coverts that are the source of the bird's name.

When the bird reached an altitude of 700 feet he began circling over his territory, saying in the language of Red-shouldered Hawks, "This is mine." But the solitary nature of his flight carried another message. It said, "There is no other." It said, "I seek. And if you also seek, here you will find."

The bird projected this message to all Red-shouldered Hawks in the range of a raptor's vision, and, in a day or two, in healthy populations there would have been eyes to read this message. In stable populations there is always a surplus stock of nonbreeding adults who may be called up to fill vacancies in a territory when they occur.

But Red-shouldered Hawks face a recruitment problem that is easy to state but difficult to explain. Except for the California population, Red-shouldered Hawks are declining. Where vacancies were once filled "promptly" (as ornithologist Arthur Cleveland Bent, writing in the 1930s, put it), now mates often go wanting, and year by year historical territories go fallow.

The reason or reasons for the decline are unknown. Speculation has focused primarily on habitat loss. Though a species that perch-hunts at woodland edge, through much of their range Red-shouldereds seem dependent upon large forest tracts. This habitat is di-

The Wind Masters

minishing, largely because of forest fragmentation caused by human endeavors — by highways, and power line cuts, and development practices. And where habitat partitioning seems to act to the detriment of Red-shouldered Hawks, it seems to benefit the larger, more versatile Red-tailed Hawk.

Perhaps the two buteos compete for territory and nest sites, perhaps for prey. Whatever the causal agent, whatever the mechanics, habitat seems to be a factor. As the forested tracts that are this forest buteo's stronghold diminish, so too does the bird.

The man had not considered selling the farm after his wife died — even though his heart was no longer in farming, even though the place harbored memories that even time could not ease. He had been born here, after all, and except for ROTC and his time in the service, he had lived here all his life. It was a good farm, a good place to live. He and his wife had been happy here.

But it had been two years since his world had been halved, and though the community was a support, it could not replace the thing that was missing in his parceled life. His friends were all married — or else they were gone. And he, who had lived for two years alone with his loss, was suddenly tired of loss, tired of being alone, and tired of the valley that seemed a vessel for his loneliness. When an agent representing a group of doctors had approached him, offering a sum that was well above the going rate for agricultural land, he had declined, but the offer was still on the table. He thought, now, that he might take it.

He carried no watch, but he could tell by the sun that it was noon and his stomach supported this guess. Because his eye had been trained for such things, because once in his life it had been a matter of life and death, he saw the Red-shouldered high overhead, circling like a fighter plane, cloaked in the glow of the sun. He noted that there was only one.

"I guess that's just you and me now," he said to the silhouette and not to the tractor, "just turning circles." Reaching for a lever, raising the plow blade, he directed the tractor out of the field and back onto the road. Once he'd gone home for lunch. But since the house was empty, he preferred, in good weather, to take his lunch outside.

"The horizon that had been empty now supported a distant bird."

The Wind Masters

The bird was still on his mind as he lurched onto the road and started making the turn for the bridge, when a patch of color caught his eye. He had to look back over his shoulder to see the car, a red sport sedan, stranded in the road. He had to look again to determine that it was indeed mired, not parked, and again before he saw the occupant who had stepped from the rear of the vehicle and was waving with both arms. It was not a car he recognized.

Overhead, the Red-shouldered Hawk saw the tractor circle and head back down the road. He noted, in uncomprehending snatches, how the tractor that he knew and the vehicle that he did not met, and how the man that he knew and the woman that he did not did likewise. The bird noted, without interest, how the man did something beneath the stranded vehicle, and how the two vehicles then moved in union, and how the man and the woman stood and talked at length.

But at this point the bird stopped taking note because the horizon that had been empty now supported a distant bird that was circling, as he was circling, and was, as he was, alone. He did not see how the man accepted a ride from the woman because the tractor, it seemed, would not start, nor later how the car lingered by the bridge.

By that time, the bird was too filled with the season and with possibility to care. It was taunting the earth and dominating the sky, screaming the joyous challenge of a Red-shouldered Hawk in spring. On the hillsides, above the bridge and below the hawks, the maples blushed and the Blue Jays mocked, but shyly.

THIRTEEN

Parabuteo unicinctus
HARRIS' HAWK

THE FEMALE WAS SITTING in the nest when the male arrived with another stick. This was a large one, nearly three feet long, but he flew without difficulty, navigating a course that steered clear of the brittle trees and wind-whipped scrub of the Texas coast. He carried the stick in his mouth.

"A stick to build a stockade with," the female more or less thought unkindly. "For four days the guy brings sticks too small and now he thinks he's Paul Bunyan."

The male gained the nest and stood on the rim looking pleased with himself, a handsome, bronze-colored raptor with the balanced proportions of a triathlete. The female, who except for her larger size was identical, did nothing.

He placed the stick at the rim of the nest, searching the face of his mate for some sign of approval but coming up short. Looking down at the stick, seeming to realize his own silly mistake, he retrieved the limb and moved it a little to the right. This didn't arouse much enthusiasm in his spouse either.

He picked the stick up again and this time offered it to her. For a moment it seemed she might refuse the offering, but finally she accepted it; took it without rising. The male hesitated momentarily, then turned and flew off in the direction he'd come. She knew he was going for another stick.

"I'll never pair-bond with a younger bird again," she promised, and not for the first time in their month-long relationship. "He means well and he's a fine hunting partner, but he's got less smarts than a day-old Red-tailed."

If the female Harris' Hawk had been among her kin nesting in Arizona instead of Texas, she might have been able to share these thoughts with one of several possible nest mates, because in parts of this southwestern raptor's range, Harris' Hawks enjoy a social structure that is almost tribal. It is common for females to be attended by two males who share all the duties and pleasures of pair bonding — nest building, copulation, nest defense, and hunting. Other, more complex interrelationships also occur in which two males and two females may share a single nest or in which an Alpha (primary) dyad or triad is assisted by a cadre of satellite helpers. These assistants may be young from a previous brood or birds recruited from afar. Their primary function is to provide food for the nestlings of the Alpha birds. In this, they perform in much the same way as the members of a wolf pack.

It is an eminently workable system and one that enjoys a high rate of success. Polyandrous Harris' raise more young, per nest, than those who are not. But the female with the oversized stick in her mouth was cut from a population of birds whose family values were more traditional in scope. As with most Texas Harris' Hawks, and most birds of prey, her relationship with her mate was monogamous — with all the benefits of its genetic certainties and its social shortcomings, too.

"Sticks when it's time for twigs," she thought, rising, fitting the stick into several possible places but finding none she liked. It's no easy task working something as contrary as a mesquite branch into a pile of like-minded sticks. Something on the branch is always catching on something else. Every potential spot has a protruding end that precludes a functional fit.

89

"*He placed the stick at the rim of the nest, searching the face of his mate for some sign of approval.*"

The Wind Masters

Once a stick does find a place, it's usually the wrong place. And trying to remove an entrenched mesquite branch from a nest of confederates is harder than trying to remove a single wire clothes hanger from a closet.

When the stick tumbled from the nest after a two-minute struggle, the female Harris' Hawk ignored it. Leaning forward she released her hold on the nest, and in a wide, arching glide that carried her halfway around the tree, she lit amid a stand of last year's coarse grass and began gathering mouthfuls.

In constructing a raptor nest, the technique is just the opposite of building a fire. Larger sticks go on first, forming the base. Modestly smaller sticks follow, then twigs, bark, and rootlets feather the interior of the shallow cup. Grass is not commonly used. This is the basic nest, and different species have their idiosyncratic variations. At one extreme are the falcons, who fashion no nest at all. At the other are Bald Eagles and Ospreys, whose nests, over many nesting seasons, may assume such grandiose proportions that they topple beneath their own weight.

The selection of a nest site is of paramount importance. The site must afford both security and proximity to food. In some places, and for some species, the shortage of suitable nesting sites is the factor limiting populations. In parts of the tree-poor high prairie, almost every suitable sapling seems to host a nesting raptor — a Swainson's Hawk, a Red-tailed Hawk, or a Ferruginous Hawk — or that great usurper of raptor nests, the Great-horned Owl.

"That damned owl," the Harris' Hawk thought, but could not say with a swatch of grass in her mouth. The curse would have been expressed as a hoarse scream and it would have been ignored by the owl whose head protruded above the nest the female had built and used the previous year. By both the hawk and the owl, it was the preferred nest, although within the hawk's territory there were three more auxiliary nests. The one nearing completion had been used as a feeding platform the previous year. This, in part, explained its size. Four feet across, and six vertical feet, a veritable tornado of sticks.

She vaulted into the air with a push and flew to the nest on rapid, paddle-shaped wings. Using her bill she added the grass to the soft inner core, then pressing her body into the cup, she snuggled in, molding the material to the fit of her body. Over the course of the next several weeks she would be spending many long hours on the nest. The attention she lavished on it now would pay her back later.

The male came in with a stick that was better proportioned than the last but, given the stage the nest was in, not terribly useful. He offered it immediately to the female, who accepted it. [Oh, a stick. How thoughtful.] Then returned it. [Hon, we've got enough sticks now. Thanks.]

The male plainly didn't get the point. He tried giving it back, and when it was refused, he placed it in front of the female and flew off.

The female closed her eyes. "He's going for another," she thought, she knew, and if a raptor can wear an expression that looks pained, then the female Harris' Hawk looked pained. "He's not even a good lover," she added, she knew, and this too was true. Copulation, like hunting, like nest building, may be anchored in innate drives — the thing we call instinct. But instinct, at least in higher life forms, is largely motivation. Execution requires skill and skill requires practice.

Nests constructed by inexperienced birds are often sloppy affairs; structurally unsound and not well placed. Copulation between inexperienced birds is likewise clumsy, amateurish; hit or miss. Nest failure, among birds nesting for the first time, is not uncommon.

"But he is a terrific hunter," she affirmed, she knew. An intelligent hunter as well as a skilled hunter. He'd proved it again that morning, when the female had pursued a jackrabbit that had taken refuge beneath a low, sprawling mesquite. The male had followed the hare on foot, flushing it toward the waiting female, who missed when the animal veered to avoid her talons — veered right into the talons of the male, who had positioned himself for just such an eventuality.

Successful hunting involves a great deal more skill than successful nest building (or successful loving), and in Harris' Hawks, cooperative hunting is the rule rather than the exception. The male was not inept, he was only young and inexperienced. Time would change that — time and practice, and it was, to the mind of the female, time. High time!

The female stood and hopped to the rim of the nest, searching for her errant mate. Leaning forward she called.

Cha cha, cha, cha, cha. It was a sound like sticks rattling in the wind. A sound well suited to this great, flat, arid land.

The Wind Masters

She stepped back into the center of the nest. Moved a stick that had been troubling her. Found a place for the most recent contribution made by her mate and then tried to remove a stick that was protruding above the rim. The stick was firmly wedged. She tried pulling from another angle but with no greater success.

Frustrated with her failure, she searched the horizon and called again. This time she saw him approaching. And this time he didn't carry a stick.

When he landed, both birds leaned forward and touched the sides of their bills like swordsmen touching blades before a duel. The female left the nest, taking a perch on the limb of a hackberry, where she waited. The male, understanding her urgency and feeling his own, did not understand the gesture or what was expected of him. He flew to the perch and lit beside her. The female left immediately.

He knew he'd blundered. He just didn't know how. And he wanted to please. He just didn't understand what was expected. Spying the stick that had fallen from the nest, he flew to retrieve it. Flying to the nest, he turned to see how this gesture might affect his mate. Her oblique response was to fly to another nearby perch.

It wasn't exactly affirmation, but it wasn't denunciation either.

He flew to the ground and tried to secure another stick, one that was still attached to its branch, one that he had tried a hundred times before — and whatever patience the female had maintained to this point disappeared.

"That's it!" she felt more than thought. And if Harris' Hawks could grasp the concept of absurdity she might have added, "This is getting ridiculous."

She flew to the ground a dozen feet beyond her attentive but still uncomprehending mate. Leaning forward in a manner that was not just suggestive but hard to mistake, she waited.

The male misread the cue anyway, and when he landed next to her, he was surprised when she turned her frustration upon him and charged. The male flew to a nearby fencepost to consider this sudden souring in the relationship, but time for contemplation was over. The female chased him off the fencepost too, and there she waited.

This time the young male got the idea. He swung around the nest tree, flew back to the perch, and lit upon her back — a little too far forward to be effective. Wings spread for balance, her tail raised and twisted to one side, they tried to make the cloacal embrace that would give meaning to weeks of courtship and give purpose to the nest. But only their loud, screaming cries commingled.

Straightening, twisting her head back and jabbing with her bill, she forced him to leave — which he did by flying around the nest tree again, then back. This time when he landed, he left nothing to be desired.

FOURTEEN

Coragyps atratus
BLACK VULTURE

THE BIRD STANDING ATOP THE ROOF of what had once
been a house interrupted his morning grooming and studied the sky.
His eyes missed little but found little to interest him — nothing, at
least, to stir the soot-colored bird from his perch prematurely.

"Beautiful morning," he assessed, feeling the warmth of the spring
sun on his feathers. "Wonderful day to find something dead to eat,"
he added — because he was a vulture, a carrion feeder, and because
he habitually talked to himself (albeit silently).

Black Vultures are *not* early risers. Not until thermals began perking and the earlier-rising Turkey Vultures had had time to sniff out
carrion would he trouble to move — despite his hunger. Black Vultures are also not strict carrion feeders. They will forage on human
garbage. They consume animal offal. They will, when opportunity
inclines, expedite the demise of sick, infirm, or newborn animals —
including the nestlings of herons and egrets. When carrion is in short
supply, they may also resort to killing healthy animals up to the size
of skunks and opossums.

But they will only attempt this in a crowd, when their shortcomings as predators have been mitigated by numbers. The feet of Black Vultures are large, sturdy, and clawed, but they cannot grasp with strength and they do not have what it takes to catch and hold prey — the mark of a true raptor. So their approach to live prey differs little from their standard scavenging technique. They grab and pull and rend with their bills, fighting among themselves for pieces, using the applied force of multiple individuals to compensate for their raptorial shortfall. The only difference is that the animal is alive — at least for a time.

Returning to his primping, the vulture began hissing a few bars of "Teddy Bear's Picnic" as he worked on the feathers of his wings. There were two reasons for this. First, he didn't know the words. Second, New World Vultures can't hum. In fact, since they lack muscles in the syrinx, the avian "voice box," they can't really vocalize. Their audio repertoire is limited to wheezes, grunts, and hisses.

Working an inner primary through his bill, he tamed a few unruly barbules, then, moving outward, took his time and got a nice even set on the six outer flight feathers, the whitish ones that formed the patch at the tip of the bird's wing.

"Sharp," he assessed, well pleased with his effort. He capped off his grooming by sending a stream of excrement down his legs — a nice touch that both soothed and cooled, a thermoregulating technique called urohydrosis. Even though it was morning and April, it was plain the day would be a hot one in coastal Georgia.

Few humans regarding the Black Vulture would concur with the bird's high self-esteem. Black Vultures are, to the eyes of most human observers, ugly, even loathsome. Squat-featured, cloaked in loose-fitting feathers, the birds look like shabby undertakers, and their dreary appearance is only enhanced by the head. It is naked — scar-tissue hard and charcoal gray. The eye, which is brown, seems emotionally detached (except, perhaps, for a touch of simple, uncomplicated madness), and the bill, which is horn colored at the tip, is cruelly hooked.

Below, in the vacant building, he heard his mate move, repositioning herself over the two one-week-old chicks whose "nest" was the open floor. Sometimes Black Vultures lay their eggs in hollow logs, sometimes in caves, sometimes in thickets; any place dark and confining seems suitable. At no time do they construct nests.

The vulture chicks were little more than naked-faced puffballs, their natal down the color of café au lait (the chicks of Turkey Vultures, by comparison, are white). They were, of course, hungry. Incubation, brooding, and feeding are tasks shared by both adults, and shift change was imminent. But first he would forage both for himself and for the young.

The bird could feel the morning's heat rising off the roof and the overgrown yard. He could see how the Turkey Vultures were no longer struggling with their flight. In the sky, tiny puffs of clouds were forming and unforming, and as he watched he saw a Turkey Vulture double-back beneath a larval cloud and vault skyward. It was time.

Leaning forward like a swimmer on the mark, the bird launched himself with a noisy series of flaps that ended in a stiff-winged glide. The bird found a thermal over what had once been a livestock pen and began to climb. Only occasionally was he forced to flap.

As he climbed, the ramshackle buildings receded, becoming just another element in the mosaic of old farms, and forests, and cypress swamps. This habitat is not universal across the Black Vulture's range, which extends from Florida's Everglades to northern New Jersey's rocky hillsides to Arizona's desert canyons. Although the birds are expanding their range northward and westward, they are still very much a southern species. In South America, the Black Vulture's range includes all but the southernmost portion of Argentina.

The vulture had no particular destination in mind. He was relying upon the carrion-finding skills of Turkey Vultures and the long-distance communication network of his kind to tip him off to the location of food. Food that may be exploited by one vulture may be exploited by another, and vultures are communal in this regard.

Setting his wings, keeping his eyes on the horizon, the bird took a course that carried him toward the coast. There seemed to be more vultures in that direction (an auspicious sign), and after filling his crop with food, he would have a good tail wind to see him home.

In flight, the vulture shed his loathsome qualities. He became instead an accomplished soaring bird — one whose abilities to soar in high places anchored him firmly within the ranks of the Wind Masters. Although not as buoyant as the Turkey Vulture, the Black Vulture habitually soars higher, in part because, unlike Turkey Vultures, Black Vultures cannot locate food by smell; also, since they are visually oriented, more altitude means more vantage.

Searching for another thermal, seeing a young Bald Eagle and two Turkey Vultures turning circles in the sky, the Black Vulture altered his course to intercept the birds. He arrived just ahead of a Wood Stork who was also searching for lift (and who was just as adept at exploiting it as the most competent raptor).

As was to be expected, the eagle, a second-year bird with a wingspan of seven and a half feet, dwarfed the other birds in the ad hoc flock. The Turkey Vulture, set upon six-foot wings, was next in size; the stork closely followed. The Black Vulture, with a wingspan of four and a half feet, was by far the runt in the pack. In shape, he most closely resembled the Turkey Vulture and the eagle — in fact, with his flat-winged profile and equally projecting head and tail, the Black Vulture seemed more a miniature eagle than a cut-down Turkey Vulture.

But, of course, in terms of avian taxonomy (the naming and ranking of birds in relation to their shared traits), the Black Vulture and the Turkey Vulture were *much* more closely allied to each other, sharing the same scientific family and, of course, the same infraorder — something that could not be said for the eagle.

The eagle is included among the diurnal raptors. The vultures, although traditionally ranked in this group, are actually, structurally allied to Wood Storks. In simple fact, vultures *are* storks. They merely look and in many respects behave like raptors — hence the long confusion.

The roots of the problem go back to the Old World, where among birds of prey there are birds that distinguish themselves as killing raptors and birds that feed largely upon carrion — the vultures. Structurally and behaviorally these two groups are nevertheless similar, having a close, common ancestry and many shared traits. Scientists studying the birds of this continent found the same pattern of killing and carrion-eating predators and assumed kinship.

98

Despite some apparent anomalies, New World Vultures, including the California Condor, the Turkey Vulture, and the Black Vulture, were assigned a "Family" ranking, the Cathartidae (the New World Vultures), and were counted among the other three familial groups of the Order Falconiformes, the Order of Birds of Prey. The other families are, of course, the Accipitridae, the group including hawks, harriers, kites, and eagles; the Pandionidae, the Osprey; and the Falconidae, whose members include the caracaras and falcons.

Doubt about the Cathartidae's standing persisted, and in the 1960s it was pointed out that New World Vultures and storks enjoy hard-to-ignore structural and behavioral similarities, among them the bare face, the absence of muscles in the syrinx, strong soaring capability, and the idiosyncratic habit of defecating down their legs for the purpose of urohydrosis. The presumed similarity between hawks and New World Vultures was not, therefore, the result of evolutionary kinship. It was caused by evolutionary convergence — the development of similar structural and behavioral adaptations to exploit similar ecological roles. Hawks and owls are another example of convergent evolution; although similar in many respects, they are very distantly related.

In 1985 evidence confirming the stork/vulture link was supplied through a biochemical technique known as DNA hybridization. The process involves the splitting of the two strands of a DNA molecule into single strands and mating these widowed strands (in the laboratory) to the split strands of the DNA of a totally different bird species, creating a hybrid DNA molecule. The more closely related two species are, the firmer the hookup between strands. The more dissimilar two birds are, the less adhesion there is in the linkage.

The way to test the hookup is to apply heat. The DNA strands of dissimilar species disassociate at low temperature. Similar species require higher temperatures to split the links between strands.

In the case of raptors and presumed raptors, it takes less heat to separate eagle/vulture hybrid strands than vulture/stork strands — hence the assumption of a closer phylogenetic link between New World Vultures and storks.

On the basis of evidence supplied via DNA hybridization, a new way of regarding the raptor/vulture/stork complex has been suggested by ornithologists Fred Sibley and John Ahlquist, who pioneered the technique in avian studies during the 1980s. In the new ranking both the hawklike birds, the Falconiformes, and the long-legged water birds whose order includes storks are now *all* included within a new, expanded order, the Ciconiiformes. Raptors are then subdivided as an infraorder, the Falconides, and New World Vultures are relegated to another, separate infraorder, the Ciconiides. These two major subdivisions are, in taxonomic fashion, further subdivided into finer categories: parvorders, superfamilies, families, subfamilies, genera, and species.

For the purposes of gaining a gross understanding of this newly clarified relation between raptors and vultures (and to illustrate the distance between them), consider a tractor-trailer, a long-haul, eighteen-wheel truck. If the entire rig were likened to the new order, Ciconiiformes, then the infraorders, Falconides (birds of prey) and Ciconiides (water birds), would represent, respectively, the cab and the trailer. Axles constitute Parvorders.

Ospreys, hawks, and eagles; falcons and caracaras would represent two separate axles on the cab. A group made up of the long-legged wading birds (herons, ibis, flamingos, storks) *and* the vultures would all fit on a single axle of the trailer. Storks and vultures would constitute a double wheel on that axle.

Roughly speaking, Bald Eagles and Black Vultures are no closer to each other than the front right tire on the cab of the rig is to the front right tire on the trailer. Black Vultures and Wood Storks, on the other hand, rest side by side on the same axle — much closer to each other than to any of the raptorial birds of prey.

How does this explain the physical similarity between hawks and vultures that suggests kinship? It doesn't. But consider how the relationship between cab and trailer might be distorted if the rig jackknifed, spinning the trailer around until it was almost parallel to the cab. In real, structural terms, the tires of the cab and trailer remain distantly associated. They only appear to be side by side.

The sun-warmed air carried the birds aloft. None needed to move a wing. Each, in time, disassociated itself on the basis of physical ability and focused ambition. The Black Vulture continued east.

It wasn't long before he saw what he was looking for — another

The Wind Masters

"The sun-warmed air carried the birds aloft.
None needed to move a wing."

101

BLACK VULTURE

Black Vulture descending earthward in a steep, steep glide that would carry him to a pig farm, a prime foraging area for Black Vultures. The white wing tips caught the sun like a mirror flashing a message for all with calculating eyes to see.

"Bingo," the vulture thought to himself.

The vulture altered his course and saw other birds do the same. As he drew closer he saw the source of the steeply gliding vulture's interest — a pile of freshly deposited rough fish surrounded by a ring of pigs. The larger animals were jammed at the center. An assortment of piglets was squeezed to the outside.

"Fish and entrails and little pig tails, that's what little vultures are made of," the vulture chanted, but silently, because vultures can't recite rhyme.

"Going to be a crowd," he thought as he started his descent, and once again the tune for "Teddy Bear's Picnic" surfaced in his mind.

Circus cyaneus

NORTHERN HARRIER

THE RED-TAILED HAWK WAS little more than a pale patch in the sky — but that was close enough in the estimate of the adult male harrier perched atop the muskrat house, too close. The slurred, down-sliding wail of his mate, huddled, like her nest, in the tall salt-marsh grass, said that she thought so too. Without hesitation the male launched himself and began climbing toward the intruder.

The angle of the bird's climb was acute, almost vertical, but the harrier's pique or his powers of flight seemed to impart immunity from the laws of physics. Wings drawn back like the blades of an arrowhead, the bird closed quickly and silently upon the circling intruder. Only when the harrier had halved the distance did he call — a staccato burst jacketed by angry, sharp-edged consonants. *Kek-kek-kek-kek-kek-kek-kek-kek-kek-kek.*

It was not a sound that left much to the imagination. The harrier was furious. His territory and his security had been violated, and right was clearly on his side.

And even though the adult Red-tailed Hawk outweighed the harrier three to one, even though the Red-tailed had all the combat advantage afforded by gravity and superior positioning, the buteo gave ground. It had tangled with harriers before and could see no fortune in engaging one of these agile, bantam-weight raptors again.

Before the harrier had stopped climbing, the Red-tailed had cocked its wings and set off in the direction of less-contested air. The harrier followed for a short time, then set his wings and began a long, arching return that would carry the bird over much of his nesting territory — about 400 square yards. To a human observer, it would have recalled the victory lap enjoyed by runners after the contest is won — and no one can say it was not. By the time the harrier regained his favorite perch, the Red-tailed Hawk was lost to sight.

There is nothing inconspicuous about an adult male harrier. Garbed in ghost-white underparts, cloaked in a hoary gray mantle, set against the tidal marshes of Delaware Bay, the bird stood out like a frosted glass statue set against a green velvet curtain. The muskrat house that offered elevation also served as a stage — one that made the hawk visible to a multitude of watchful eyes.

This suited the bird's interest and needs. Sentries are supposed to be obvious. How else are intruders to know that they are being intruders? It was the female, mantling a nest and its seven eggs somewhere upon the open marsh, that needed the cryptic colors of concealment, needed them as much as and maybe more than she needed a sentry's protection. The reason is simple. Female harriers cannot always depend upon the timely intercession of their mates. As principal providers, male harriers spend a great deal of time away from the nest and their foraging flights may take them ten miles or more from the nest. These distances and absences are necessary because, in many places, including the Delaware Bay shore, harriers are polygamous. Males may tend two, three, and sometimes more mates simultaneously, and the food demands inherent in multiple parenting make foraging a full-time job.

The sentinel male had two mates. One was the bird whose wailed warning had caused him to intercept the Red-tailed. She was his first and favorite mate, the one who garnered most of the male's attention, and her nest was closer to his perch. The other female's nest was

The Wind Masters

just over two hundred yards away. As the slighted member of the trio she was, mostly out of necessity, something of a nag. In order to secure food, she was often obliged to leave the nest and solicit, driving her mate to the performance of his duties with whining, repetitive *eeeehah* calls. She had even been reduced to pirating food meant for the other female once or twice.

Earlier in the season, there had been a third female in the male's harem, a second-year bird that was not an adult but was, like many birds of prey, still capable of producing young. Her first nest had been poorly elevated, and when a recent storm-swelled tide had flooded the marsh, it had swamped. It was a typical amateur's mistake. Following the disaster, the young female's allegiance had shifted to another male in an adjacent territory, and, for some reason, the male atop the muskrat house had not contested the change.

Maybe the sentinel male had calculated the loss of prey that would result from the flood and realized that the demands of three broods would overtax his foraging abilities, putting all the vessels of his genetic inheritance at risk. Maybe the other male had simply out-rivaled him, or perhaps he had simply tired of three mates.

The young female's first nest had been near the border separating the respective nesting territories — a long, sparse line of saltbush that cut across open marsh. Her nest was actually closer to the nest of the second male's primary mate than it was to either nest of her harem associates. What's more, earlier in the season, she had been courted by and copulated with both males. Perhaps her allegiance had never been resolute, and neither, apparently, had her first mate's.

From his perch, the male could see the neighboring male soaring high over the marsh, gaining altitude steadily. Like an athlete on the bench studying the performance of another star performer, the sentinel male watched the other bird's movements.

Watched, as the performing male's easy, ascending spirals reached their peak.

Saw, how the wings narrowed to points and how the avian projectile plunged to earth.

Observed, the fluid, wing-flapping, trancelike dive of the other bird.

Felt, just as he saw, the gravity-cheating set of the wings on the

NORTHERN HARRIER

other bird as he pulled out of the dive and clambered for the sky, climbing almost as fast as he had fallen.

Then, at the height of the bird's climb, the performing male executed a perfect somersault before plunging earthward again, to begin the sequence again . . . and again . . . and again . . . a ritual of courtship that has been aptly and colorfully described as "sky dancing."

"Nice technique," the sentinel male noted. "Little tight on the roll, but the stoop is poetry; dead-on accurate."

Earlier in the season, this courtship display would have touched off copycat displays in adjacent territories across the marsh — as performing males tried to outshine their rivals. But with most females well into their incubation, all it aroused in the sentinel male was proprietary interest.

It wasn't that the sky-dancing bird was throwing a glove. His rejuvenated spirit was linked to the ritual process of re-nesting; to the young female, whose clutch of replacement eggs was still incomplete. Sky dancing, among harriers, is much more a signal of prowess and sexual readiness than a mechanism for marking out territory.

But sky dancing is a way of saying: "I'm the hottest, spunkiest, most capable (and most eager) harrier in the marsh," which is just another way of saying: "I'm better than you are." And the capitulation of the young female had let a measure of uncertainty creep into what had been a stable domestic *and* geopolitical situation. There is a certain amount of risk in leaving your territory for extended periods when you have a neighbor who thinks he is "better than you are."

Abruptly the sentinel bird left his perch, caught a thermal, and spiraled skyward. There was no hurry, as there had been when the Red-tailed had crowded the bird's territory, and no need to climb for elevation when nature's elevator was operational. A hundred yards over the marsh, the bird set his wings and started a long, calculated glide that would take him into the other bird's territory.

Territoriality, the establishment of spatial boundaries that are aggressively defended against other members of the same species, is the rule among most raptors. They do this to protect the resources upon which their lives and the lives of their mates and nestlings depend. They do this to ensure the genetic integrity of their offspring.

The Wind Masters

Different species have different needs and use different standards when establishing territories, but the size of a bird's territory is generally determined by the space it feels it needs to defend itself, its mate, and its young (a nesting territory) and the amount of area it needs to hunt successfully (the hunting territory, or home range). Often the two are not the same.

Harriers are rather social by the standards of most raptors, incredibly tolerant of others in their own species. In winter, they often roost communally, within yards of each other, and fan out during the day to forage. Even during the breeding season, there is a great deal of overlap in hunting areas, and two or more pairs may work the same patch of marsh or field *simultaneously*. As such, the area around a nest that the bird will defend against intrusion by another harrier may be quite small, several hundred square yards — not much more space than is required to play American football.

The advantage inherent in defending a small territory is considerable. The less territory you have, the less time you must allocate to defending it and the more time that can be directed toward other duties — such as hunting. This consolidated defense strategy is one of the factors most often cited to explain polygamy among harriers. Less time defending and more time hunting means greater opportunity to succeed with multiple mates and young.

Territoriality benefits local populations of a species as well as individuals within a population. The establishment of territories is an effective check against overpopulation. Where adequate nest sites are not a factor and where prey is generally abundant, birds of prey distribute themselves evenly across an area — they spread out! By establishing and defending territories, they ensure that more birds will not move into an area than the area can support.

It is roughly similar to a human community that gauges the resources of an area (recharge capacity of the aquifer, highway infrastructure, capacity of the sewage treatment facility) and establishes zoning regulations. The difference is that human communities are planned and raptor "communities" simply result. There is another difference. In most human populations, transgressions relating to spatial distribution and proprietary boundaries are enforced through courts of law. In birds of prey, territorial defense is the responsibility of the birds themselves.

The male harrier's course carried him over the primary nest, eliciting a cry from his Alpha mate — a reminder of duties (which he had, in fact, not forgotten). Her cries elicited several cries from his other mate (whom he also had not forgotten, just neglected). The male harrier continued toward the saltbush hedge and the location of the old nest that had been flooded, toward the female and her new nest, with its incomplete clutch of eggs, that lay beyond. As he approached the demarcation line, he was not at all surprised when the other male ended his display and moved to intercept him. In fact, he anticipated it.

The flight of a Northern Harrier is one of the most fluid movements in nature. Slight of body, set upon sliverlike wings, the birds can move with the grace of wind-borne leaves. This is how the bird was flying when he approached the other bird's territory. But when a harrier has need for speed, it can race like a greyhound and twist and turn like a mink. This is how the defending male moved to intercept him.

The sentinel male, who was now a transgressing male, altered his course slightly to skirt the edge of his rival's territory, but he did not retreat. The bird whose territory was being tested responded not by directing an attack upon his neighbor, although it is not uncommon for harriers to pursue and grapple with each other when the issue of territory lies between them. Instead, the resident bird assumed the position of escort a dozen feet behind and slightly below the trespasser. In the body language of harriers, the message was clear: "You cannot land here."

But for reasons that can only be guessed at, the intruding male ignored the diplomatic warning. He raised his wings high above his back, dropped his legs, and began to circle. The defending bird adopted the same posture and circled in kind. They circled this way twice, and then both lit upon the marsh, very close to the invisible boundary line, each bird facing the territory of the other.

In the body language of harriers, this message too was clear: "This is mine."

The birds faced each other this way for about ten seconds, heads held high, chests thrown forward, yellow eye to yellow eye. They had

The Wind Masters

"*The resident bird assumed the position of escort a dozen feet behind and slightly below the trespasser.*"

NORTHERN HARRIER

done this earlier in the year, before their territories had been defined; had even grappled earlier in the year when territories were still disputed. But the bird who had been sentinel and was now trespasser had been moved to raise the issue again.

Ten more seconds elapsed, during which time neither bird moved and neither vocalized, but from the territory of the trespassing male the voice of the favored female reached their ears. It was the call she had used before, the food-begging call, but this time it was more petulant, more insistent — almost nagging.

Several more seconds passed, and then very slowly, very deliberately, almost as gracefully as a harrier flies, the intruding male opened his wings, letting the tips trail in the grass, and then leaning forward he dropped his head, almost touching the grass. It was an astonishing gesture, but one that did not leave much to the imagination. A gesture that looked for all the world like, and was in every way imaginable, a bow.

He did this twice before his neighbor acknowledged the gesture by returning it in kind. In the language of harriers it was as good as saying, "Yes, that is yours."

It was going to be a long breeding season, one filled with demands and uncertainties. The border between the males' respective territories would not be one of them. When the birds departed, they flew in opposite directions, one to attend his new mate, the other to fend for his established ones. Neither looked back.

The Wind Masters

SIXTEEN

Falco peregrinus
PEREGRINE FALCON

PAST UKUTLIK BEND, the cliffs rise four hundred feet on a side and the current quickens, recalling the river as it was in its upper reaches, deep in the foothills of the Brooks Range. Past this point, there are other bluffs, but none so tall or so expansive that all the cliff-nesting raptors of the Arctic can find grudging accommodations.

A Rough-legged Hawk here. A Gyrfalcon, a Peregrine, or a raven there. Nothing like the shale- and coal-seamed high-rise that mantles the bend, stained red by the lichen that grows in the places that nesting raptors anoint with mute.

Above one such stain on the west-facing bluff, a bird huddled in the remains of an old stick nest. Her helmeted head and back were the color of slate in rain, and she might have passed for a stone, so still did she lie, except that her head moved in measured jerks, betraying her presence and her unease.

She searched for danger. She searched for diversion. But most of all she searched the skies for her mate, who had been gone over an hour. It was late afternoon, feeding time for Peregrine Falcons. Although she was as capable a hunter, she had dispatched her mate with a nagging series of food-begging commands and was eager for his return.

In the division of Peregrine duties, it is the female who assumes the balance of tasks associated with the nest. Males, who are smaller and less devoted defenders, do most of the hunting. It is a practical and proven arrangement and yet . . .

And yet she was a Tundra Peregrine, a bird whose concept of distance is defined by the horizon and whose migrations vault hemispheres; a bird whose very name, *Falco peregrinus,* means "wanderer." She had incubated her clutch for twenty days, would for twelve to fourteen more, and she was restless. Restless, and perhaps in some vague fashion envious of a mate who used his morning and evening hunts to exercise and wander, leaving her to tend and defend.

Only her eyes were free to roam, and these she turned upon her world — upon the river swollen with snowmelt; upon the opposing cliff face, the distant snowcapped peaks of the Brooks Range, and the thunderstorms that raced across the rolling tundra of Alaska's North Slope, touching the lives of creatures ten, twenty, fifty miles away.

She could not see these creatures. Even the eyes of a Peregrine have their limits. But from the ledge she could easily observe more immediate neighbors. These included hovering jaegers and towering longspurs, nesting Glaucous-winged Gulls and grazing caribou.

Across the river and a mile down, she could see the female Gyrfalcon tending her young on the lichen-stained ledge that was the pair's traditional stronghold, and half a mile away, on her side of the river, she observed the raven kids cutting up around their abundantly whitewashed nest. But these were not merely neighbors. They were rivals who earned the special scrutiny that goes with such a distinction.

As odd as it might seem for a falcon to nest in proximity to other predators, the Peregrine's closest neighbor would arouse greater curiosity. Below the Peregrine, less than a stone's toss away, was the huddled form of the Canada Goose. Beneath the goose was her clutch of five eggs.

If the Peregrine seemed like stone, the goose seemed like part of the

The Wind Masters

cliff itself, so quietly did she lie and so closely did she resemble the shale piled around her. But the bird's quiescence was not due to any fear of the falcon. Indeed, the goose had selected this site for her nest to take advantage of the Peregrine's defensive umbrella and the curious, benign indifference Peregrines seem to bestow upon birds nesting close to their nest sites — even birds that Peregrines call prey.

Nest defense was a factor governing the falcon's choice of sites, too. Although in some parts of the world Peregrines will lay their eggs in open tree cavities or prebuilt nests, they are for the most part cliff-nesting birds, a strategy to foil four-legged predators and some two-legged ones. The ideal site is a ledge no less than two feet wide that overlooks and is protected by water below. The ledge should be high enough to deter predators who might climb, and low enough to foil advances from above. An overhang above the nest is a bonus both for deflecting hungry eyes and for sheltering birds from the elements.

Though serviceable, the site occupied by the female Peregrine was not ideal. The cliff was sloped, not vertical, and therefore not unassailable. A determined, sure-footed predator could find ways of reaching the nest ledge from above. Nor did the ledge offer much more than meager protection from the elements — in fact it was quite open above.

The nest sites of choice were the ones occupied by the ravens and the Gyrfalcons. Both were large, roomy enclaves cut deep into the face of the cliff. Both offered a southern exposure for greater warmth and comfort, and the rock face above and below was sheer, affording maximum protection. The only way to reach either site was to fly.

The ledge occupied by the ravens would have been the Peregrine's first choice — had, in fact, been the ledge that she and her mate had occupied the previous year. But ravens are permanent residents in the Arctic, and their prize for enduring winter hardships is first choice of nest sites. When the Peregrines had returned from their winter quarters during the first week in May, they'd found the ravens already on eggs. After two days of open hostilities, the falcons relocated to their present site — adopting the nest built by Rough-legged Hawks the previous year.

PEREGRINE FALCON

Peregrines, like other falcons, do not build their own nests. They lay their eggs in rocky "scrapes" or appropriate strategic nests. Indeed, it has been suggested that Peregrines and Rough-legged Hawks may have a commensal relation (much as Gyrfalcons and ravens do) — that in the Arctic, Peregrines may depend upon Rough-legged Hawks to build and periodically maintain nests.

Whatever the truth, it is plain that in the game of musical ledges, early-nesting species such as the raven and the Gyrfalcon are the winners. Rough-legged Hawks who arrive while Gyrfalcons are already incubating eggs run third. Peregrines, who do not reach the Arctic until May, must settle (or, more appropriately, resettle). Among the various ledges traded and occupied along the bend, the one now occupied by the Peregrines was the least favored.

"*Ehhhh*," the male Rough-legged, whose nest was around the bend upriver, called. "*Eh-ehh-eh*." It was a breathy call with a peevish, whiny quality, and frequently given. As a species, Rough-leggeds are more sensitive to intrusion than Peregrines, venting their displeasure by calling and taking wing even when potential predators are far from the nest. But the Peregrine's Rough-legged neighbors were touchy even by Rough-legged standards. Though the Peregrine could not see the buteos from her ledge, she could tell that the male had taken to the air and that he was agitated about something. What she could not tell was what. What she needed was her mate.

"*Ny-aaaah?*" she called. It was a wild call, a no-nonsense call, and although there was something about it that was plaintive, the call had another, harsher quality that cut like flint. The cry was not uttered with vehemence or volume, yet the sound of it carried, eliciting chortles from the raven kids and engendering stillness that was utter in the goose.

What the cry did not return was an echo from the Peregrine's mate.

Abruptly the female Peregrine rose from the pile of sticks, stepped away from the precious contents, and launched herself over the river. On wings that rippled like water over stone, she swung away from the cliff, gaining elevation and the vantage she needed.

Half a mile upstream, on a ledge just above a six-foot seam of coal, the pale head of the female Rough-legged showed above her nest.

The Wind Masters

The buteo was vigilant but immobile. Her mate was circling above the cliff but he was not calling now, and short of climbing to his height, the female Peregrine had no way of telling what had disturbed him and whether it had any bearing on her nest. Of her own mate, there was still no sign.

The temptation was strong to mount the skies, to satisfy her curiosity and unleash her wanderlust. But her obligation to the nest was stronger. She returned to the ledge, unfurling feet that were shockingly large, and landed.

They could kill by constriction, these celebrated feet. They could kill with an open-fisted blow delivered with thunderbolt speed. Though not as large, or as powerful, as the feet of a Golden Eagle, they were more universally feared. Peregrine Falcons are nearly cosmopolitan in their distribution, nesting on every continent but Antarctica. Nothing that falls between the size of a longspur and an eagle and flies in open country is beyond the reach of those talons. No creature, whatever its size, encroaches upon a Peregrine's nest site with impunity.

The reason lay at her feet. Four eggs, nearly as large as hen's eggs but more oval and ornate. In shape they described the elliptical orbit of a planet; in color they were cream based but so liberally spattered with burnt reds and browns that these overlying colors coalesced and eclipsed it. They had the luster of polished stone, the texture of a lover's touch, and a warm glow that recalled life itself.

In a word, they were beautiful; among the most beautiful eggs in the world of birds. They begged a hand to reach out and enfold them — and many a hand could not resist that temptation. In the acquisitive Victorian era, Peregrine eggs were a favorite among egg collectors, who plundered nests. In England, where egg collecting was particularly popular, the detrimental practice continued into the twentieth century, only to be eclipsed by a more insidious and widespread threat to Peregrine populations, DDT. The chemical agent, introduced in the 1940s to control insect pests, had an unanticipated side effect. It inhibited a female Peregrine's ability to transport and redistribute calcium, the principal component of eggshells. The result was eggs that were 15 to 20 percent thinner-shelled than shell tolerances engineered by nature.

The poisoned eggs were still beautiful. But they were also dysfunc-

PEREGRINE FALCON

tional — breaking beneath the weight of incubating birds; causing widespread reproductive failure in Peregrine populations over most of Europe, Asia, and North America. By 1970, Arctic Peregrine populations had been trimmed to between one quarter and one third of historical numbers. One subspecies, the one that occupied North America east of the Mississippi, called the eastern *anatum,* was extirpated. In North America, only the *pealei,* a large, dark race localized to islands and archipelagoes off the coast of Alaska and British Columbia, passed through the DDT era relatively unscathed.

Though many raptor species felt the effects of DDT, it was the demise of the Peregrine Falcon that won broad public sympathy and became an environmental flash point. The evidence and the outcry was enough to put an end to the widespread use of DDT. Declining populations stabilized, then began to rebound. Over the course of two decades, as the residue and the effects of DDT disappeared, Peregrine populations over most of North America returned to historical norms. And the eggs whose fame was once founded in their beauty assumed a new poignancy. They became symbols of the unmindful hardships our species levels against the environment.

But the bird, staring down at her clutch of eggs, knew none of this. Her affinity for her eggs had nothing to do with their aesthetic value or environmental ethics. It had to do with the instincts of a parent, and they were absolute.

With a degree of care that some might feel was at odds with the nature ascribed to one of the earth's most celebrated predators, the falcon reached into the nest with an open bill, carefully turning the eggs to reapportion her body's warmth. Toes closed, feet limp, she shuffled carefully into position above her clutch. Then, spreading her feathers, she settled, pressing the bare warmth of her incubation patches over the oval vessels; enfolding all in a blanket of feathers. The bird was hardly settled before a cannonade of stones cascaded down the slope, falling all around her.

Her departure this time was neither careful nor leisurely, and had the eggs not been cradled in a nest it is possible that two would have been thrown. But caution came second to need, and even before the bird turned to confront whatever creature dared threaten the sanctity of her eggs, she was venting her anger and screaming her intent.

The Wind Masters

NY-YAHH, NY-YAHH, NY-YAHH . . . NY-YAHH, NY-YAHH, NY-YAHH . . . The sound sent shock waves across the tundra and echoed from the cliffs. It brought every head within earshot to bear upon the ledge, and not a few creatures, great and small, changed their plans and adjusted their paths so as not to draw the misdirected heat of the female Peregrine in her rage.

Even the wolverine, picking a path down the cliff, was not unmoved, but anger in another creature only seems to fuel the ferocity that burns in these Arctic predators. Even grizzly bears give ground to these large, ill-tempered members of the weasel clan, and nothing that lives in the Arctic and whose existence is tangential to the earth can claim immunity from the creature's marauding — not even Peregrines.

The wolverine was not even trying to be circumspect. It knew where the falcon's nest was. It wanted the eggs. And it had fairly calculated their price. When the bird stopped calling, the weasel tensed but did not retreat.

So much has been written about Peregrines and their prowess that truth would seem fancy's victim, but just the opposite seems true. Peregrines *are* fast, so fast that in this age of precision measurement their maximum speed is still conjectural; so fast that whether the blow delivered by a diving falcon is dealt with an open or closed fist remains a matter of debate.

In level flight, a Peregrine cruises at 40 to 55 miles per hour, and when hunting, with wings beating 4.4 beats per second, a Peregrine can attain speeds of 70 miles per hour — as fast as a sprinting cheetah; nearly fast enough to catch a racing pigeon or a fleeing Canvasback duck. But it is when Peregrines place themselves at odds with the horizon, wedding the power of their wings to the earth's gravitational pull, that speeds are attained that fairly exceed the sum of their parts.

A German researcher, E. Hantge, clocked the "Wander-falcens" at 170 miles per hour in a dive angled at 30 degrees, and 220 miles per hour at angles measuring 45 degrees. In 1975 an English researcher, D. A. Orton, used mathematical estimates pertaining to the speed of free-falling objects to calculate that a Peregrine, in a vertical stoop of 5,000 feet, would be able to attain a maximum velocity of 230 to 240 miles per hour.

PEREGRINE FALCON

Mathematical ceilings notwithstanding, analysis of film footage taken of a stooping Peregrine during World War II led researchers at the Naval Research Laboratory in England to the startling conclusion that the feathered missile was traveling at 275 miles per hour, 35 miles per hour faster than Orton's projected ceiling. As astonishing as this figure is, it pales next to the touted, but unverifiable, estimate made by a fighter pilot in 1930. While diving on a flock of ducks (for practice), the pilot was startled to see a Peregrine pass him at a speed that he judged to be *twice* his plane's air speed of 173 miles per hour.

The enraged female Peregrine was not traveling at 350 miles per hour as she stooped on the wolverine. She was not traveling at 275 or 240 or even 170 miles per hour. There was no time to "tower up" above her adversary, to add the advantage of gravity's lever to her plunge. When she fell, she had less than 200 vertical feet to apply to her rage; enough to propel her 1,200 grams to a speed just shy of 80 miles per hour. Much of this force she transferred to her feet just before she struck, raking the weasel's back with an opened-taloned hind claw.

The animal buckled, snarled, and turned, confronting empty air. It lost its footing and skidded down the slope, tumbling into a stand of stunted willows that checked its fall — and saved it from another punishing blow. All that touched it of the Peregrine's second stoop was a rush of air.

NY-AHHH, NY-AHHH, NY-AHHH. The bird called as she climbed and fell silent as she stooped. She cried more loudly when the animal moved and softly when the wolverine paused to consider its route. Using the scant vegetation to its best advantage, the marauder continued down the slope, past the nest ledge, seeking paths that avoided open areas. It was not afraid of the Peregrine, but it had felt the bird's talons and was respectful.

The wolverine stopped its descent and began working across the ridge, just below the nest ledge, moving from boulders to bushes like an infantryman dodging fire. The Peregrine, moving like a possessed pendulum, continued her barrage and her harangue. But for all her anger, and for all her skill, she could not dislodge the creature.

Had the male Peregrine been present, the defense would have been formidable — maybe even daunting enough to deter a wolverine.

The Wind Masters

"The Peregrine, moving like a possessed pendulum, continued her barrage and her harangue."

PEREGRINE FALCON

But the male was not present and the contest would have ended badly, should have ended badly for the Peregrine but for one thing. Chance — one of the most determining principles in the universe. The goose, who would be stone, did not smell like stone and, when approached too closely, did not act like stone.

The goose held her ground until the wolverine was barely ten feet away, then flushed, leaving her eggs to the predator, who quickly opted to accept easy fortune in place of hard potentials. Besides, goose eggs are larger than Peregrine eggs and more numerous.

Though the wolverine's ambition was redirected, the Peregrine's anger was not. The bird continued to harass the wolverine, forcing the mammal to take the eggs, one at a time, to a protective stand of willows in order to consume them. When the animal finally fled, the last egg cradled in its mouth, the Peregrine harassed it all the way to the top of the cliff and onto open tundra beyond. Only when she was certain that the animal was not coming back did she return to the ledge.

Swinging wide over the river, the bird lit upon the ledge but did not settle upon the eggs, not immediately, and the reason lay at her feet. Chance is fickle. It gives and it takes, and although it saved the Peregrine's nest from the wolverine, it did so at a price.

The Peregrine studied the dislodged rock that chance had directed into the old Rough-legged nest, but it took several minutes for her to understand that the two eggs she had dislodged upon departure were all that was left of her clutch. The other two lay beneath the stone.

"Ny-ahhh!" she cried, but the hard edge of it was blunted by what may have been puzzlement but what sounded like loss. Maneuvering the surviving eggs with an open bill, she tried to compose herself over them in the usual way but discovered that the stone prevented this. After several futile efforts she turned 90 degrees, and by facing the cliff she found that she could mantle the eggs by crowding the rock. It wasn't comfortable, it wasn't ideal, but it was workable.

She was on the eggs only a short time when her mate returned, a partially eaten golden plover in his talons. He circled in front of the nest, inviting the female to take the meal, and when she did not he flew to an outcropping where he deposited the morsel before flying to the sentinel perch he favored.

The Wind Masters

SEVENTEEN

Buteo swainsoni
SWAINSON'S HAWK

TO THE WEST, a hundred miles away, the Rockies still slumbered beneath a mantle of snow — would until July. But on the prairies of southeastern Wyoming the land was unsheathed by spring. Over the short grass, there were towering longspurs, tail-chasing Lark Buntings, and gyrating Common Nighthawks, whipped into a lustful frenzy.

Hardly a corner fencepost could be found that did not host a meadowlark puffed up with song. Scarcely a pebbly pasture existed that was not the launch pad for some hormonally fueled Horned Lark. And above the horizon there wasn't a cloud that did not frame a hovering Ferruginous, a soaring Swainson's, or a kiting Red-tailed Hawk.

Everywhere the prairies were in motion. Everywhere there was energy unleashed by spring. Everywhere there was courtship and territory establishment, and nest building, and creatures struggling to meet the demands of parenting.

Everywhere, that is, except along a low, sloping hillside that fell without hurry toward a cottonwood-rimmed creek. Scattered on the grassy expanse were forty or so creatures, wandering freely, who seemed as unaffected by the frenetic pace surrounding them as were the distant peaks.

From a distance, the assemblage might have been mistaken for cattle. Some were rufous, some black, some piebald; all were foraging belly-deep in the grass of a rain-rich spring. But cattle graze with their heads down, not raised as these creatures did, and cattle navigate on four legs, not two.

If these discordant points were not undermining enough, the creatures in the curious assemblage displayed one more attribute that easily distinguished them from cattle. They could fly.

First one, then another of the birds lofted into the air, displaying long, tapered wings and a buoyant flight. Soon nearly all were airborne, gathering like leaves in invisible streams and coalescing into swirling eddies over friendly updrafts.

Some birds toyed with the wind, facing down zephyrs with the steadfastness of kites. Others tilted at imperceptible windmills or sparred with shadows. There didn't seem to be any purpose to the ad hoc air show. No objective, no ambition, no reason at all.

Unless the birds were simply surrendering to some vestigial restlessness left over from their long migration north. After nearly two months of travel and 6,000 air miles, flying must be a difficult habit to break.

The air show lasted several minutes before the birds seemed to tire of their sport. In ones and twos and informal threes, they left the hilltop, glided back down the slope, and settled once more upon the grass, where they began foraging again.

The birds were subadult Swainson's Hawks, a slim-winged buteo species of open places, hatched the previous summer, who were only now approaching the end of the first year of their lives. Though together, they were not a flock, not in the strict sense of the word. Their association was informal, circumstantial, not coordinated. They were not migrating together and it was only the food hidden in the grass that was their bond, nothing more.

The birds were not only independent among their union, they

The Wind Masters

"First one, then another of the birds lofted into the air, displaying long, tapered wings and a buoyant flight."

SWAINSON'S HAWK

were innocent bystanders to the frenetic pace around them. Unlike the other prairie birds, the subadult Swainson's Hawks had no territories to establish or defend. They had no mates to court, no nests to build, no eggs to incubate, and no nestlings to feed. In short, they had no obligations. Swainson's Hawks, like most buteos, do not mate until the end of their second year.

Most of the birds had only just arrived, late stragglers in the wake of one of North America's most stupendous animal migrations. Some had been raised locally by adult birds who were already deep in the process of bringing forth new young. Others were birds who heralded from farther north — from Alberta, Saskatchewan, even Alaska and the Yukon — birds whose migratory ambition had failed to keep pace with the advancing seasons.

But they had this much in common. They were the same age and they had just completed (or nearly completed) the great pan-hemispheric leap of their species, from the plains of North America to the plains of Argentina and back — a distance of more than 12,000 miles. They had been gone eight months, four of them spent traveling.

And they were home. Kids who had gone away to a very tough school, passed their finals, and seen something of the world.

They had made it! Now they could relax and turn their attention to feeding, and resting, and preening. They could stand apart from the activity all around them and enjoy the halcyon days of subadult raptors.

It was an odd-looking assemblage; disheveled, almost ragtag. Most in the group were white-spotted-on-brown above; brown-spotted-on-white below. Not a few had facial markings that might be likened to chocolate dribblings extending down the cheeks onto chests. All were missing feathers; all showed vast amounts of wear on those feathers that remained — from sun, and lice, and abrasion, and flight. No two birds were alike.

There was one individual, working the lower parts of the slope, that was as dark as the bottom of a thunderhead and another, perched higher up on the hillside, that looked as though she had been held by the tail and dipped to her middle in whitewash. Morphologically and geographically between these extremes was a third, piebald bird a-dorned with salt-and-pepper speckling on his head, breast, and wings.

Also present among the pack was a bird whose facial pattern bore an uncanny likeness to that of a young Tundra Peregrine and another bird that had all the characteristics of an adult, light-form Swainson's Hawk — except for his bib. The bib was streaked, not full; as spare as an adolescent's beard.

Plumage was not the only characteristic distinguishing one bird from the next. The hunting techniques employed were equally idiosyncratic.

Thunderhead, the dark one, was a still hunter. His technique was to find some grassy bowl and then stand, head high, searching the ground around. Prey sighted, he'd strut to the spot, bury his bill into the grass, and extract his prize. Transferring his victim to a raised foot, he'd halve the morsel and swallow it in two head-tossing installments.

Whitewash liked to fly. Sometimes she'd spot something and fly to intercept it. Sometimes she would fly to a spot and hope to find something worth intercepting. Much of the time she just watched Thunderhead. After he scored, she'd hurry over to the spot and hope to capitalize on his fortune.

Piebald was a stalker, a careful, one-foot-ahead-of-the-other stalker. His head swayed like a cobra, sometimes high, sometimes low. When he pounced, he landed like a fighting cock, an unorthodox feet-firster. Most of the other feeding birds secured prey with their bills.

There were two things about the assemblage that might have drawn an observer's interest, and the first related to the hidden prey. It wasn't large, not even rodent size, but whatever it was, it must have been incredibly abundant. How else to explain the ad hoc concentration of birds (not to mention the bulging crops)?

Second, among the feeding birds, there was a noticeable absence of adults — although there were many in the area. Chestnut-bibbed light-form birds, rufous-bellied intermediates, even one absolutely stunning dark-form Swainson's that looked like cinnamon-dusted chocolate. Yet none of the resident adults seemed interested in the bounty of prey being reaped by the subadults. Even more surprising, no adults came over to drive the nonbreeding birds away — even though the sloping meadow overlapped the home range of two different pairs of Swainson's Hawks and was less than half a mile from the nest of a Ferruginous Hawk.

The question of prey and the apparent generation gap are related.

The subadult birds were feeding on grasshoppers, big, succulent, yellow-winged beauties that went down with a crunch. Crickets and grasshoppers constitute the principal prey of all Swainson's Hawks, both old and young, after the breeding season and during the austral winter. But during the breeding season, adult Swainson's Hawks switch to warm-blooded and reptilian prey. It's not a matter of taste, it is a matter of expedience.

Grasshoppers are easy to catch but expensive to transport. It's one thing to stand hock-deep in small prey and gobble them at leisure. It's quite another to capture an insect and transport it some distance to the mouths of hungry nestlings. For a bird the size of a Swainson's Hawk it is far, far more energy efficient for adults to transport full-size prey such as ground squirrels, birds, and snakes to nestlings than it is to truck insect morsels. The energy payload is greater; the frequency of shuttle flights mercifully reduced. What's more, noninsect prey is richer in calcium — a food resource that fast-growing young Swainson's Hawks need to survive.

Eschewing insect prey offers another advantage to Swainson's Hawks — a social one. Insofar as subadult birds utilize a different food base than adults, the respective age groups avoid competing with each other. This means that adults do not have to waste time and energy defending their borders from a bunch of happy-go-lucky subadults enjoying their halcyon days. Subadults, for their part, avoid harassment and do not jeopardize the nesting success of their kind.

Competition between adult Swainson's Hawks and other prairie raptors is not so neatly avoided. In fact, there is considerable overlap in the mammalian prey utilized by nesting Swainson's Hawks and Red-tailed and Ferruginous Hawks, and territorial disputes between these two early-nesting species and late-arriving Swainson's Hawks are real and earnest.

Interspecific competition extends to nest sites, as well as to ranges, because on the tree-poor prairies just finding a place to nest can be a difficult charge. Swainson's Hawks meet this challenge by being fairly open-minded with regard to site selection. Isolated trees, telephone poles, low bushes, abandoned structures, even sloping ground will serve if nothing else is available — or if a Great-horned Owl has appropriated the site of choice.

With a breeding range extending 3,000 miles, from northern Mexico to Arctic Canada, it is impossible to establish a universal arrival date. But the great masses of Swainson's Hawks, some 300,000 to 500,000 strong, come boiling out of Mexico and across the Rio Grande in late March and early April. From there they fan out across the continent to complete a journey that may yet be measured in hours, days, or weeks.

As soon as adults arrive, territory establishment and courtship begin. Neighboring Red-taileds, whose territories have bulged past old and mutually acceptable boundaries, must be put in their place. Bonds between mates, even mates of long standing, must be renewed — females must be charmed by acrobatic, tight circling displays that end at the nest tree.

But it doesn't end at the nest tree. In fact it only begins. With sticks and twigs piled below the crown and a cup lined; with bark and grasses to cradle one to four eggs; with an incubation period (shared by both adults) that lasts about 35 days. Then the demands of parenting begin.

Males do most of the hunting; females tend nest and young until the chicks are about half grown. The young are voracious! Fast growing and fiercely competitive. Even when prey is abundant, young Swainson's Hawks commonly outstrip the capacity of parents to provide it, and fratricide is not uncommon.

Survivors fledge after 40 days, though adults may not be totally free of their food demands for several more weeks. Until the first cold weather of autumn casts a spell of torpidity upon grasshoppers. Until the swelled hosts of Swainson's Hawks leave this hemisphere behind and travel 6,000 miles to the next one.

On the hillside the subadult birds continued to feed, oblivious to all but the insects at their feet. A few birds with bulging crops had moved down the slope to preen, but they made no move to go and had no intention of going. Why should they?

Everywhere around them there were birds in motion, but for all the young Swainson's Hawks knew or cared it might have been a light year away. A span too far to see or contemplate from the vantage of a halcyon summer.

SWAINSON'S HAWK

DAS

EIGHTEEN

Rostrhamus sociabilis
SNAIL KITE

AFTER WAITING SEVERAL MINUTES, the bird attacked the shell again. Pip, pip, pip. Pip, pip . . . pip, pip, *pip*. Exhausted by its efforts, the bird rested. In the course of the bird's life, as a member of a species specially adapted to defeat the armor of the freshwater apple snail, it would open many shells. But this shell, the one that had sheltered the hatchling since its inception, was the first calcium threshold it had to cross. Until it freed itself there would be no other challenges and no life as a Snail Kite.

The egg was situated in the shallow cup of a large, bulky stick nest, the most prominent feature in the ocean of grass that is known as Water Conservation Area 3A, a vestigial part of Florida's once vast Everglades. More or less supported by willow saplings, the nest was an *average* height of eight feet above the water — a curious way to express this bit of datum, but nevertheless apt. Seven and a half feet above the water on one side, eight and a half on the other, the structure was listing badly.

Adding her weight to the unsteady structure was the incubating female, a harrier-sized, brown-backed bird with a mottled breast and an Osprey-like mask across her eyes. Her mate, a slate gray beauty, perched nearby. She was one year old, he was four. This was her first nesting attempt; his second. As breeders, neither was what might be called an old hand.

Pip, pip, pip. When the bumping began again, the female rose, stepped carefully to the elevated side of the nest, and looked down, regarding her clutch.

There were two eggs pressed against the side of the nest. Before the thunderstorm that had raked the wetlands and undermined the structure, there had been three. This number is about standard for Snail Kites, and for birds of prey in general. Two, three, or four eggs constitutes the average clutch size for raptors. Larger species such as the eagles and the vultures lay fewer eggs, sometimes only one; smaller species such as the smaller accipiters and, particularly, the harrier species lay more (up to seven eggs in the case of the Northern Harrier).

The kite's eggs were large relative to the size of the bird, almost Ping-Pong-ball size, and this too is typical of birds of prey. The shells were also fairly conventional — smooth and not particularly glossy; more broadly spherical than classically ovate. The background color was cream, generously spattered with dark and pale brown markings.

These traits also fairly typify the eggs of North American birds of prey, although egg size varies as widely as species, and colors range from the unblemished white of the eggs of the Northern Harrier to the dark, pattern-dominated reddish brown of those of the White-tailed Kite.

Appropriately, only the eggs of the three vulture species diverge radically from generalizations regarding shape. The eggs of Turkey Vultures and Black Vultures are undeniably ovate; the egg of the California Condor is clearly elliptical.

The shell itself was hard and resilient; its principal component, calcium carbonate structurally knit by calcite crystals. For 28 days the shell and its contents had provided all that a developing raptor required — food from a lipid-rich yoke, water bound up in the cushioning ocean of albumen, protection from the outside world's threats.

It was a brilliant and beautiful biosphere, one of nature's most

"When the bumping began again, the female rose."

The Wind Masters

perfect creations. All that it lacked was the incubating warmth of the parent (which in the case of Snail Kites both parents are quick to provide) and room to grow.

As incubation periods go, the 28 days needed to transform a kite embryo into a kite nestling is highly precocious. Among North American raptors, the average incubation period is closer to 32 days, and some species require considerably more time to develop. Goshawk eggs, for example, may take 38 days to hatch; Golden Eagles, as many as 45. California Condors remain in the egg for 57 days.

But in half the time it takes a condor to hatch, the developing Snail Kite had outgrown its shell. Two days earlier, in its 26th day of development, it had begun the task of getting free — its first act of volition and the onset of the greatest struggle this nonmigratory species would confront in its life.

Lying on its left side, feet and belly contouring the more tapered end of the shell, the hatchling brought its head up between its body and its right wing — a feat of contortion that would have earned it the envy of Houdini. Forcing its bill forward, the kite punctured the shell's inner membrane at the egg's blunt end. Above this barrier was an open space, an air cell. Beyond it, the shell.

Having gained a little room to maneuver, the bird began attacking the shell, butting it weakly but repeatedly not with the tip of its bill but with a small, hard cutting tool located right at the curve of the upper mandible. This tiny eggshell opener, known as the egg tooth, is lost soon after hatching.

After repeated strikes, the chick breached the shell. It wasn't exactly a hole, nothing as dramatic as that. It wasn't even an open seam. It was just a "pip." A craggy raised bump in what had been a blemish-free shell. But it was a weakening in the structure. The first. The chick celebrated the milestone by resting.

After a short pause, the chick began pushing with its legs and pivoting counterclockwise within the shell. Bringing its head forward once more, it began driving the hard edge of the egg tooth against the shell . . . again . . . and again until its efforts were rewarded with another pip next to the first one. After another rest, another adjustment, and more effort, there were three pips — the beginnings of a fracture line whose projected path would circumnavigate the shell.

Now, nearly two days later, the line was nearly complete and the chick was nearly exhausted. Now, if it could muster the strength, it was time to confront the challenges that Snail Kites face beyond the protective confines of the shell.

There were many challenges. And for this federally designated Endangered Species perhaps more than those faced by the nestlings of most other North American raptors.

As a nestling, the young kite would have to confront, avoid, or surmount the usual challenges imposed upon raptor nestlings — predation by raccoons, harriers, rat snakes, and Boat-tailed Grackles (the usual enemies of Snail Kites); starvation resulting from periodic shortages of prey.

One of the most common causes of nest failure in Snail Kites stems from the nest itself. Though these structures are reliable enough when placed on a firm foundation — a sturdy tree such as the pond apple or melaleuca — in times of drought kites often build their nests in improper vegetation in order to be closer to snail-bearing water. Willow saplings, cattails, saw grass, and bulrush don't have the fiber to stand up to the weight of a Snail Kite's nest — particularly when storms rake the glades. The plants bend or break, the nests tip, eggs and young land in the water and perish.

It has been estimated that less than one kite nest in three fledges young, with most failures occurring during incubation (and even before incubation). When the foundation supporting a nest is bad, it doesn't take long to show.

To combat this uncommonly high rate of failure, Snail Kites have adopted a breeding strategy tailored to the situation and the nomadic nature of the species. When young are about half grown (two weeks old), one of the adults may desert to start a new nest with a new mate. What makes this strategy possible is an unusually long nesting season (January through August in North America) and an abundance of easily procured food. When water conditions are favorable, keeping young kites in apple snails is a one-parent operation — provided you know how (which all adults do). Catching apple snails is, after all, what being a Snail Kite is all about.

From a springy perch, on bowed, paddle-shaped wings, hunting kites sally forth in search of their prey. Coursing just above the marsh, the flight is languid, unhurried — but then it doesn't take much alacrity to catch a snail. It does, however, take good vision. The brown-shelled apple snail may be found on plant stems just above or below the surface, where it breathes and filter-feeds.

Stalling, swooping toward some open lead among the rush, the birds unleash their extremely long legs, plucking the fist-sized mollusks from the water with a single foot. The birds can also perch-hunt with equally fine results.

The snail is carried to a nearby perch, where kites apply their patented brand of epicurean artistry. The utensils that make it possible for a Snail Kite to dine on snails include large, snail-enfolding feet and a grotesquely long, exquisitely curved snail-picking bill.

Pinning the snail to the branch with the opening facing away from the perch, the bird begins working its bill around the edge of the operculum, prying the protective trapdoor off with a twist. Turning the shell so that the door faces up and the spiraling curve of the shell complements the curve of the bird's bill, the kite drives its curved upper mandible deep into the snail's sanctum, shearing the columellar muscle, separating the animal from its armor. Then, using both mandibles, the kite extracts the snail, letting the shell fall.

It's a subtle bit of artistry that young kites must learn by trial and error. It is also as limiting in its application as it is specialized. So supremely adapted are Snail Kites to their prey that they can scarcely feed upon anything else, which is fine — so long as there are apple snails to be had. Unfortunately for the kites, this is not always the case.

Drought is probably the greatest natural enemy that kites face. When these periodic cycles parch the land, standing water recedes and snails respond by burying themselves in the mud — beyond the sight-guided reach of the Snail Kite's feet. In 1980 there were 651 Snail Kites in Florida. Following two years of drought, this number was halved — which represents a dramatic decrease but does not even approach the dearth of birds found (or *not* found) in 1972. In 1972, following another period of drought, North America's Snail Kite population was a mere 65 individuals.

Cold fronts and their associated northwest winds also hamper the efforts of kites to secure prey. Falling temperatures prompt snails to go deep, and water turbidity, caused by high winds, deflects the eyes of kites. It has been suggested that the persistent threat of cold fronts, more than the range of the apple snail, is the factor setting the northern limit of the Snail Kite's range — the headwaters of St. Johns River (about halfway up the peninsula).

From this northern limit, the birds range south to Lake Okeechobee, the Everglades, Cuba, southern Mexico, and Central and South America to northeastern Argentina. In the United States, which is to say Florida, the birds range widely over the southern portion of the peninsula when water levels are high. When drought closes its fist over the land, the birds withdraw to deep-water refuges such as Lake Okeechobee and the managed water conservation areas that lie west of and between West Palm Beach and Miami. Within these jurisdictional sanctuaries, where water levels rise and fall to balance the needs of Florida's agriculture industry, its human population, and its wildlife, the birds seek shelter and wait for the rains.

Shelter — or, more accurately, its constricting confines — was still very much on the mind of the hatchling. Rested, though hardly recovered, the kite-to-be was more than ready to trade the world it knew for the uncertainties of the one it did not. The bird made a concerted attempt to straighten itself. Not just turn, straighten — putting pressure on the weakened end of the shell.

It wouldn't budge.

Rest again, try again. The shell continued to defeat the bird's best, feeble efforts. Having no other recourse, the hatchling returned once more to the laborious task of pipping. Then, some time later, it tried straightening again.

This time, the shell separated, slightly.

It took another hour for the chick to struggle free — nearly two days since he'd first set bill to shell. In a few short months, it would take the bird less than two minutes to open the shell of an apple snail.

That was a challenge for another day. Right now the naked, blind creature who looked a good deal more like shelled prey than a predatory bird was going to rest.

NINETEEN

Buteo regalis
FERRUGINOUS HAWK

THE LANDSCAPE ROLLED like an ocean and looked like an Easter basket. Where the prairie wasn't green with grass it was yellow with flowers, and it had no boundary but the sky, which was blue and cloud-pocked.

It was the first real spring that the prairies had seen in five years, the first spring after the drought had broken. During that time the creatures that call the grasslands their home had adjusted their lives to conform to harsh realities. Some by leaving, some by dying, many by limiting the number of offspring they would add to a resource-strapped world — by withholding life itself.

The Ferruginous Hawk had been one of these. She and her mate of three seasons had raised only two birds two summers ago and had chosen not to breed at all the previous year. When conditions are favorable, meaning when prey is abundant, Ferruginous Hawks may fledge three or four young in a season. But there is no fortune in trying to raise a family when most, and maybe all, will not survive.

The rains had finally come, late in the summer; too late for Ferruginous Hawks to breed. The summer rains had been followed by a good snow cover, followed by a long, wet spring. And now life had returned to the prairies with a pent-up vengeance. From her perch, on a limb set close to the trunk of a creekside cottonwood, she could see McCown's Longspurs displaying, Horned Larks tail-chasing, Swainson's Hawks soaring, pronghorn antelope grazing, and cattle being moved out to range.

These were the boundaries of her horizon, and, added to the nest with four young, and the mate who was hunting, they defined the limits of the bird's life — and would for several more weeks. Her mate's role was to provide food for her and for the young. Her role was to protect and tend.

It wasn't as exciting as hunting. It didn't give her the freedom to mount the skies and watch the landscape change, and race jackrabbits with her shadow. But it was better than incubating eggs.

"Anything is better than incubating," she had long ago concluded. For 33 days, and with very little relief by her mate, she had done precisely that.

One of the chicks moved in the nest, a gangly ball of matted fluff and half-grown feathers — but she did not trouble to look. The chick was three weeks old, nearly half grown — old enough not to require *constant* attention.

If it had been raining hard, as it had rained the night before, or if it had been cold, it would have been another matter. The female would have been off her perch and on the nest, brooding. But it wasn't cold and it wasn't raining and she had nothing to do until her mate returned with prey, and so she sat. Half hidden among the branches of the tree. Staring out at a horizon that was as removed from her life as the spring that had deepened into summer.

"It's better than incubating," she affirmed again, but her eyes were set on the horizon.

Little of the female was visible, which was intentional. Despite their size and weight, which make the Ferruginous Hawks the largest buteos in North America; despite their bearing, which is, as the name *Buteo regalis* suggests, lordly, *regal*, Ferruginous Hawks do not accommodate intrusion, particularly human intrusion, easily or well. If the bird had wanted to advertise her presence, she would have taken a higher perch. One that would have shown off her most striking features to great advantage — the great white chest and underparts; the broad white tail with the rose-blushed tip. This is how her mate had asserted himself to her and rival males so many weeks ago — before the snow had ceased to fall; before spring on the prairies was hardly a rumor.

It was in April that they had surrendered to the hormones racing through their blood and brought closure to their courtship. It had begun with visits to several potential nest sites, auxiliary sites. It ended with a food pass. Their bond had been renewed on the ground near the most favored site, and renewed repeatedly, over the course of many days.

The site of choice was an old one, the same one the birds had used two years earlier, usurped from a Red-tailed Hawk. With trees at a premium, usurpation is common. The nest was set in a cottonwood, the smallest of five trees but nevertheless the tree farthest from the rutted land that served the ranching community. The structure was forty feet off the ground, fifteen feet below the topmost branches, and flanked by full-leaved limbs that were flush with the rim of the nest. From the road it was invisible.

It had taken a week to bring the nest to a state of readiness. The male had furnished most of the heavy material — the sticks and roots and the clods of dried cow manure that would form a cushioning layer to cradle the female and her eggs during incubation. She had molded the material to her taste. When it was finished, the nest measured forty inches across and nearly that much deep, which is very large, even by the standards of buteos. But despite its size it was very difficult to see.

Ferruginous Hawks choose to nest in trees when and where they are available. But the lives of these birds are wedded to dry open grasslands — in fact, the breeding range of this geographically restricted buteo virtually defines the prairies of the American West. Where suitable habitat, a surfeit of prey, and no trees exist, Ferruginous Hawks make do — with cliffs and buttes, if available; hillsides

FERRUGINOUS HAWK

and dirt banks and boulders and even haystacks where no better nesting substrate is found.

Ground nests are, of course, more vulnerable than elevated nests and, for many Ferruginous Hawks, anonymity the first and final defense, particularly where human intruders are concerned. Once a nest has been compromised, once a female has been disturbed during laying or incubation, abandonment occurs in one out of three cases.

She continued her vigilance, turning her head to regard the great sameness of the landscape and to search for the small inconsistencies that might mean danger. Little of what she saw concerned her — except for the horsemen accompanying the cattle, and these concerned her only a little. The ranchers were still more than two miles away — too far to see her or her nest. Still . . .

A commotion over the rim of a distant rise caught her eye. A pair of Killdeer were circling and calling loudly, and several Red-winged Blackbirds were bobbing over some nemesis whose identity was concealed by the horizon. Some instinct told her it was her mate before her eyes confirmed it.

She watched as his form drew near. A large, pale, long-winged bird with rust-colored shoulders, a blush-colored tail, and white patches bleeding through the wings. The russet leggings, feathered right down to the feet that held the ground squirrel, were not visible.

He flew to a rocky outcropping a hundred yards from the nest, deposited his prey, and started away before she reached the site. These prey exchanges were old hat now, more perfunctory than even their mating had ever been.

The female gained the outcropping and the ground squirrel; saw that the male had already eaten the head — which she would have done herself, had he left it. Then, pushing off with her feet and down with her wings, she flew directly to the nest and proceeded to tear the animal into strips — offering them to whatever young mouth was more incessantly open.

A single ground squirrel does not go far among four voracious young. In less than five minutes, the rodent was gone and she was back on her perch. Nothing had changed — except that the male had taken a position on a hillside and the cattle were closer. As she watched, he moved his bill along the length of a wing. Several loos-

The Wind Masters

"She continued her vigilance, turning her head to regard the great sameness of the landscape and to search for the small inconsistencies that might mean danger."

139

FERRUGINOUS HAWK

ened feathers joined those already scattered on the hillside. After several minutes, he started off again. He would have stayed longer. But there were ground squirrels to kill and nestlings to be fed. The routine of domestic life wasn't much fun for a male Ferruginous Hawk either.

She watched him go, noting the gaps in his wings where 5th and 6th primaries had been shed and new ones had yet to grow. In February, when their courtship had started, his plumage had still been fresh and full. Now . . .

She was conscious, suddenly, of her own worn plumage — the frayed flight feathers; the scapulars whose colors had been bleached by the sun. Absently, she began to work her bill over several unruly tail feathers, feathers that had been damaged during incubation. Occasionally, she noted the progress of the cattle and the horsemen. They were about a mile and a half away now. They seemed to be following the course of the stream.

The silhouette of a Golden Eagle crested the horizon, distracting the female Ferruginous and commanding her attention. But the eagle knew where the hawk's territorial boundaries lay and had no reason to force them. The eagle, an adult, had responsibilities, too, and no cause to test the good nature of neighbors.

There were other birds up and flying. Red-tailed Hawks that kited and hovered over the prairie. Chestnut-chested Swainson's Hawks that waltzed across the sky on uplifted wings. A Prairie Falcon that interrupted its spiral flight and started the long, shallow stoop that would bring it to grasstop level and end with some unwary creature's demise — and if not this time, then the next time or the time after that. It was all just a matter of time.

The Ferruginous Hawk sat for half an hour, during which time nothing changed (except that the cattle got closer). She dropped to the nest and picked up the remains of the ground squirrel and discarded them. The young begged to be fed. What else was new?

Another half hour passed and finally her restlessness got the better of her. Abruptly she left the nest.

Her inclinations were to fly, to mount a thermal and soar, but her duties would not allow this. Instead she flew to a knoll some two

The Wind Masters

hundred yards from the nest. Though set lower than the tree, it offered a fine view of the nest and the horizon — and the approaching cattle, which were beginning to make her nervous. She arched her back. Muted. Scratched the side of her head with a raised talon. Yawned. And then, turning into the sun, sat like some chest-heavy, anvil-headed penguin.

It wasn't flight and it wasn't freedom, but it was good to be away from the nest even for a little while. The young would be in the nest for three more weeks. After that, there would be four more weeks when the young were entirely or partially dependent upon the adults for food. Then . . .

The bird turned her attention back upon the horizon — the horizon that was as far away as the past and as alluring as the future. She saw her mate when he was a frail spot on the horizon and waited patiently for him to close the distance, drop his prey, and retreat to his place on the hill.

With a glance toward the cattle, who were now quite close, she went to retrieve the ground squirrel whose own monotonous routine had been interrupted by her mate. She carried it to the tree and proceeded once again to tear the animal into strips, offering pieces to mouths that never closed.

It was all routine — except for the cattle who were only several hundred yards away and who had smelled the water lying beneath the trees. The two horsemen, who were looking forward to a place to sit that didn't bounce and was out of the sun, were eager, too.

FERRUGINOUS HAWK

TWENTY

Buteo nitidus
GRAY HAWK

A PHAINOPEPLA WHISTLED A WARNING. The song of a Varied Bunting was brought up short. The barred shadow, the color of the monsoon rain falling upon not-so-distant Mexico, materialized in the heat-warped air and lit upon a rocky outcropping, then turned facing the arroyo. *His* arroyo.

The bird, an adult Gray Hawk, was hunting — but not out of hunger and not for himself. True, the bird *was* hungry. The small whip-tailed lizard the bird had consumed early in the day had not satisfied him for long (in fact, had not satisfied him even then). But it was late June, the hot height of the breeding season in Arizona. Hungry or not, the male Gray Hawk had other mouths to feed besides his own and they were not far away.

Along a seasonal stream corridor that had yet to feel the touch of summer rains, nestled in a stick nest high in the canopy of the tallest cottonwood were three downy young. *His* young.

Perched on the rim of the nest, offering what shade she could from the afternoon sun, was the Gray Hawk's mate. She resembled the male in all respects, including hunger, because as the nestlings' principal guardian, she would not leave the nest to forage until the birds were older. She, too, was dependent upon the male for prey.

Were all three of the nestlings to live to maturity, they would represent a measurable increase in the number of Gray Hawks nesting in the United States. The adult male and his mate were among 60 pairs believed to breed north of Mexico — almost all of which nest along the Santa Cruz and San Pedro river drainage systems in Arizona. Among all of North America's raptors, only the California Condor and the Hook-billed Kite are more numerically depressed.

Before the face of the Southwest was altered by cutting, grazing, and agriculture, Gray Hawks were more common and more widespread. The small buteos flourished in the riparian woodlands and mature mesquite forests of Arizona, New Mexico, and Texas. Though the species' range extends southward into tropical and subtropical regions, Gray Hawks are not birds of wet, humid forests. The bird is a dry woodland hawk, at home among the branches that mantle a sparse understory, where lizards skitter among the leaves and where the bird's rapid acceleration and quick reflexes are used to best advantage.

If Gray Hawks in the United States choose nest sites along riparian corridors, it is because this is where suitable nest trees are found, not because the birds are dependent upon aquatic-based prey (as is the case with the Common Black-Hawk). Prey for the Gray Hawk includes birds, small mammals, but especially reptiles — an agile prey for an agile raptor.

If the male's hunting perch was located on an open, desert hillside, overlooking an overgrazed floodplain studded with immature mesquite, the selection was a matter of compromise — the avian equivalent of pushing the envelope. The male Gray Hawk would have preferred to hunt in the more mature woodlands that lined the banks of Sonoita Creek, but the bird did not have that choice.

There were, along the course of Sonoita Creek, four established nesting pairs of Gray Hawks — the carrying capacity of the habitat. The best the three-year-old male and his mate could find was the small cottonwood grove flanking the seasonal stream and the over-grown ranch land surrounding it.

There was sufficient prey within the bird's territory. That was not the problem. The problem was the vegetation that covered much of the area — young, ground-hugging mesquite and acacia with branches so interlocked that even a bird as agile as a Gray Hawk could not pass through. A lack of strategic hunting perches, and the long approaches to sighted prey that this obligated, also hampered the bird, reducing his success. That the bird continued to provide prey to his nestlings and his mate was a credit to his determination and skill.

That the bird was hungry attested to the limit of his determination and skill.

Suddenly the hawk left his perch. One moment he was there — just another ash-colored stone set against the desiccated hillside. The next he was gone, transformed into a buteo projectile propelled by wing beats that were almost too quick to count. The bird's course was direct, determined, no-nonsense. It seemed certain that prey had been sighted and that some small creature was about to die — but this was not the case.

The hawk sped across the arroyo, melting into the heat that made objects ripple and the horizon dance. Just before the bird and the opposing hillside became one, the Gray Hawk opened his wings, climbed slightly, stalled, and lit upon another outcropping. The bird was merely changing perches. All or nothing is simply the approach Gray Hawks apply to their endeavors — whether this means defending territory, capturing prey, or changing position.

From his new vantage, bathed in searing sunlight, it was difficult to believe that the bird was not, by choice, a desert predator. His neutral plumage might have been woven of heat waves themselves, so well did the bird meld into the dry environment. His rapid, open-country dash seemed ideally suited for life as an open-country predator.

But anatomical evidence says otherwise. Gray Hawks are slim and compact, not rangy like many open-country birds. The wings are short, shorter than the wings of a Broad-winged Hawk, a forest buteo (and one to which the Gray Hawk bears more than a passing resemblance), and the tail is fairly long — proportionately longer than, say, the tail of a Red-tailed Hawk.

The Wind Masters

"*From his new vantage, bathed in searing sunlight, it was difficult
to believe that the bird was not, by choice, a desert predator.*"

GRAY HAWK

Short wings and long tails are traits that are well suited to woodland raptors such as the accipiters, the Sharp-shinned Hawk, Cooper's Hawk, and Goshawk. In fact, not long ago the Gray Hawk was called the Mexican Goshawk — a colorful name recalling at once the bird's geographic orientation and its enthusiastic, all-or-nothing hunting style (not to mention the buteo's superficial likeness to an adult Northern Goshawk).

But the species name was changed to Gray Hawk, and while the new name is less evocative, it is no less apt. The adult bird *is* gray — ash gray above, gray barred below; the wings seem fashioned from slate. Only the tail breaks with the uniformity of this pattern. Two black bands, two white. When backlit, the white bands gleam and the paler flight feathers of soaring birds glow with translucent light.

Set against their ashen backdrop, the yellow cere, gape, and feet of Gray Hawks are striking, even ornate. But the bird's most arresting feature may well be its eyes. Large and brown, intelligent and alert, they seem at once both fierce and benign.

At the moment, the Gray Hawk's eyes were trained upon the latticework of branches and shadows that covered the floodplain. The young were hungry. The knowledge preyed upon the bird's mind.

Though the afternoon heat was still fierce, the sun's greatest strength was past. Very soon the temperature-sensitive creatures that sit out the worst of the day's heat would begin to stir. On the banks of the arroyo, beneath the branches of a young mesquite, an Abert's Towhee was foraging. But the towhee might just as well have stood behind a wall of talon-proof armor. No dash, no matter how spirited, could penetrate those branches, and Gray Hawks are short in the shank, not blessed with a dinner-table reach.

On the hillside, ground squirrels were going about the business of being ground squirrels — foraging freely (calculating constantly their speed against the Gray Hawk's speed and how this related to their distance from burrows and safety). The Gray Hawk paid them little mind.

Finally, the hawk saw what he was looking for. A spiny lizard — a large one, almost eight inches long, about a hundred feet down the slope. It was lying in the shade beneath a young mesquite whose

The Wind Masters

branches just *might* allow a Gray Hawk passage. The reptile had betrayed itself by moving to avoid the rays of the circling sun. A ground squirrel whistled a warning but the bird was already *gone*. Eyes fixed. Talons drawn tight against barred underparts. Wings moving with clipped precision. Each succeeding wing beat seemed quicker than the last until the wings were moving so fast that in the heat-softened air, downstroke and upstroke merged and became one. Halfway to the lizard, the hawk set his wings, trading a measure of speed for the greater advantage of surprise. When he reached the outer edge of branches, the hawk was forced to brake with wings and tail to slip beneath the curtain of branches. This cost him more speed, and worse than this, it cost the bird the element of surprise.

The lizard was alerted by the sudden movement, but it reacted to the hawk's shadow, dodging right, when it should have sped straight away; dodging right . . . right into the open talons of the shadow-colored bird. Even if the lizard had been a better guesser, it's possible that its celebration might have been short-lived. Gray Hawks are tenacious, tough in the clinches. The hawk may well have pursued the reptile on foot, and though lizards are fast, Gray Hawks are fast and determined.

The bird did not wait until the lizard had stopped thrashing before reaching down, pulling off a strip of flesh, and bolting it down. He downed one more morsel, thoughtfully pried from the less meaty fore end of the creature, then springing into the air, he flew toward the creek, caught a buffeting thermal, and began to climb. Another bird might have turned circles, letting the superheated air bear it aloft — but not a Gray Hawk, and particularly not a Gray Hawk pushing the envelope in marginal habitat.

Impatient with the speed of his climb, lending support with his wings, the Gray Hawk gained altitude swiftly, leveled off when he knew that he could clear the ridge, then turned his thoughts and his wings toward the distant grove of trees.

There, beneath the canopy, were young in a nest and they were his. Understanding nothing about marginal habitats and the hardships this imposed, they were aware of nothing but their hunger, which would, in small measure, soon be assuaged.

Elanus caeruleus
WHITE-TAILED KITE

INBOUND HIGHWAY 101 COMMUTER TRAFFIC went from congested to sluggish to stop-and-go. Kenn Johnson's irritation index level mounted in perfect, inverse accord — from peeved to irritated to angry. When traffic stopped utterly, his composure blew completely.

"What now?" the forty-year-old engineer demanded of the bumper of the minivan ahead of him. But the bumper, being a back bumper, was as uninformed as Kenn.

He tried opening the window and peering around the van to gain some conceptual understanding of the source of the delay. But the van wouldn't cooperate here either.

"Maybe it's just the construction," he thought, he hoped. But as time moved on and the traffic did not, he knew that the morning's delay wasn't construction related.

"Well, maybe it's just a fender-bender southbound and a little rubbernecking," he theorized. But this projection proved optimistic too. The three lanes of traffic didn't budge.

"It *can't* be another accident," he pleaded, gripping the steering wheel of his sport sedan with both hands, pushing himself back into the seat. Involuntarily he thought of his wife, who was still his wife, but only because the papers hadn't gone through yet.

"I love you," she had said at the door, the day she'd left.

"But you hate it here," he said, before she could, completing the litany that had been repeated so often during the last year of their marriage that it had assumed the properties of a chant. A Wyoming native, accustomed to and craving open, unpeopled landscapes, she'd felt persecuted in the crowded Bay Area and she particularly hated the traffic and the development that spawned it. She felt that she might have stayed if they'd just stop destroying what was left — but they didn't, and in the end she didn't either.

"It will get better," he'd pleaded.

"I can't wait," she'd replied. "I've got to get on with my life."

"What about the boys?" he'd ask.

"They've never known any other place and they're almost grown. They'll need you now, not me."

So after seventeen years and two children they separated. She, to pursue a new career (and another man) in Colorado. He had stayed, keeping the sons, his job, and their home.

"I can't believe this," he said both of his shattered marriage, his stalled life, and the stalled traffic. With an abrupt, angry gesture he threw the car into neutral . . . then park . . . then he turned the engine off. Totally disgusted, he looked around, seeking support and consolation from fellow commuters, flanking condos under construction, and the grassy, gold-colored hillside beyond.

Fifty feet away, sitting atop a young eucalyptus tree beside the highway, was a gull-sized hawk with gentle eyes. Despite his striking white underparts, contrasting black shoulders, and bobbing, eye-catching tail, the commuter's eyes went right past the bird. He never saw it.

The bird was a White-tailed Kite, a medium-size raptor of grassy plains, marsh edges, and open, irrigated country. It nests from Washington State south to Baja California; in pockets in Arizona and New Mexico; then coastally in Texas, Louisiana, Mississippi, and south-central Florida. The bird was formerly much more widespread and

WHITE-TAILED KITE

more abundant — occupying most of the lower Great Plains. The demise of this beautiful and unwary raptor was so nearly complete that it is something of a miracle that the bird survived at all to share the roadside with Kenn Johnson.

By the mid-1800s overgrazing had nearly eliminated California's lush, coastal grasslands and the California vole (*Microtus californicus*) they harbored — both the habitat and the prey favored by the bird. Then the great wetlands of the Great Bear state were diked and drained for conversion into agricultural cropland — a land-use practice that further reduced grassland habitat and voles (and White-tailed Kites).

The trusting demeanor of the bird was no asset. It made the kite an easy target for senseless, widespread shooting. Also contributing to the bird's demise was the beauty of its eggs. Creamy white but overlaid with, and almost eclipsed by, warm reddish brown tones, they were, in the estimates of many, the most beautiful eggs on the planet and were coveted by egg collectors.

By the 1930s North America's White-tailed Kite populations were almost gone; the bird seemed destined for extinction. In 1957 the California legislature accorded the bird the same protection privileges accorded the California Condor. But even before legislative action, the birds had begun making a comeback — first in California and soon throughout other parts of their historical range. Ironically, the recovery (like the decline) was rooted in changing human land-use practices.

Cattle and crops don't mix. The fences erected to separate the two industries created miles of linear, grass-edged foraging habitat for voles and house mice (*Mus musculus*) — man's perennial companion (and an alternate prey item for the beleaguered kite).

Shortly after World War II, the kites were the principal beneficiaries of another, purely human institution — land speculation. People hoping to one day turn a profit on their investment bought up grazing land that was then allowed to go fallow while prices caught up to dreams. Further unwitting allies of the kites were wealthy individuals who wanted privacy, not profit, and for whom fallow land was testimony to their wealth. Newly purchased and once overgrazed

The Wind Masters

land was allowed to restore itself. Drought and wildfire kept it open. Kites flourished.

Two other developments may have helped the kite to expand its numbers and its range — irrigation farming and interstate highways. The practice of drawing water from aquifers or distant reservoirs to irrigate desert areas may have opened whole new regions to kites. Deserts in Arizona and New Mexico, once too dry to support the rodents White-tailed Kites need, became rodent-friendly and kite-friendly with the addition of water.

Despite a reduction in shootings and the virtual elimination of egg collecting, human persecution of birds of prey has not been eliminated. The recreational shooting of raptors, particularly in rural areas, continues. But gunplay is actively discouraged in places where large numbers of people flourish — in places such as the suburbs of San Francisco, for example. While much of urban and suburban habitat is unsuitable for kites, there is, in these vast metropolitan labyrinths, a network of transportation links called highways that support flanking grass borders — ideal hunting habitat for kites. Few people are going to cross a busy California interstate just to harm a bird (provided they even notice it).

When the headline news station Kenn had been listening to started running through the world events lineup for the third time, he started punching buttons, trying to find a program he could live with. Given his present mood, this would have been asking a lot from a mere radio station, and it is not surprising that his ambitions fell short.

He switched from FM to AM and things just got worse. He went back to FM and, just to fuel his frustration, went back through all the selections again. No soap — and no movement in the line of cars either.

Looking for reprieve, looking for anything, he looked out the passenger-side window, managing, somehow, not to see the kite that was hovering thirty feet from the car. This was understandable, since Kenn had no ken of birds, but it was also unfortunate because he was desperate for diversion and because he did enjoy beautiful things. White-tailed Kites are, if nothing else, beautiful.

Male or female, the underparts are stark white; the back, soft gray; the head and tail, silver blushed. The head itself is stocky, almost neckless; the bill, petite and black. But the bird's most arresting feature is its eye — an almond-shaped shadow surrounding a ruby orb. It made the bird hovering near the roadside seem either baleful or infinitely wise. The gape, straight cut and neutral, supported neither conclusion (or both) — not that the stranded commuter cared. He still hadn't seen the bird.

Even a grand funk of frustration, such as Kenn Johnson was experiencing, cannot be sustained forever. His irritation finally gave way to resignation and resignation to boredom. Stifling a yawn, he rifled through his stock of cassette tapes, running his fingers over the efforts of recording artists whose names haven't appeared on the charts since the invention of the microchip.

He settled on a movie soundtrack, realizing sadly and almost immediately that it was a film he and his wife had seen when they'd first met. He switched to a self-help tape, then tried the radio again, then turned the system off, preferring silence.

The kite preferred silence as well, but neither the noise, nor the traffic, nor the fussing antics of the stalled commuters distracted the bird from his purpose. Like work-bound Kenn Johnson, he had a job to do and, also like Kenn Johnson, there were kids to be fed — four in all. Two males, two females, all recently fledged and, except for a light rufous streaking on the head and breast, looking very much like the adult.

In addition to a partiality to silence, Kenn and the kite had several other things in common. They more or less shared the same activity period. Kites are active for the first several hours after dawn and from midafternoon until evening — about the same time humans engage in that great mobile relocation called "commuting." During the middle portion of the day, kites become more or less sedentary (just like working humans).

Kenn and the kite were also single parents. Both had been left by their respective spouses and for much the same reason: to get on with their lives. Particularly when prey is abundant, it is common for female White-tailed Kites to desert newly fledged young and seek out a new mate. The old mate is left to tend the fledglings until they can fend for themselves, a period that may take several months.

The Wind Masters

From the standpoint of both the species' population and the individual's genetic imperative, this post-fledging shuffle makes wonderful sense. It maximizes the survival of existing young, increases a female's productivity, and offers both males *and* females more opportunities to spread their genetic inheritance around. With so many free-floating females actively seeking mates, male birds, once divested of young, are also able to recruit a new female into their territory and nest again.

Kenn let his eye roam over the ranking of cars caught up in the tie-up — admiring those that carried a bigger sticker price than his vehicle, dismissing those that did not. He tried to make eye contact with the woman whose face was reflected in the rear-view mirror but garnered nothing more for his effort than a short, dismissive glance.

Desperate for anything to do, he reached over for his briefcase and drew out several papers relating to a morning meeting he was supposed to attend. He put them back after several minutes when it occurred to him that unless traffic began to move he would never make that meeting. His life, pure and simple, was stalled.

The kite stopped hovering. Flapping stiffly but languidly, he climbed to fifty feet, flying parallel to the line of cars. Had Kenn seen him, he might have concluded that the bird was some sort of gull. But the hunched shoulders, graceful uplifting of the wings, and rocking flight would have contradicted this conclusion. When he stopped abruptly in midair and began hovering again, all resemblance to gulls ended.

Body angled, yellow legs dangling, wings flicking down and forward, the bird held his place in the sky, but all his concentration was on the grass below. Sometimes he paused in flight, wings raised in a sharply accented V, playing wind against gravity, holding both forces to a draw. It was a beautiful and stunning display of precision flying. The bird's secondaries rippled in the wind like the tense arms of a gymnast executing moves on the rings. And when the bird did move forward, his gliding flight came as easily as an expelled breath.

This happened several times, and after each "kite" the bird moved up the line, starting the procedure again. Kenn, who was still playing the old rear-view mirror game with the female commuter, missed it.

The fourth time the bird stopped, he stooped, but it wasn't the

"Sometimes he paused in flight, playing wind against gravity, holding both forces to a draw."

plunging stoop of a Short-tailed Hawk or the fluid swoop of a White-tailed Hawk. It was more nearly a parachuting glide. The bird fell like a dream, and only when he was ten feet off the ground did the fall seem to catch up to real time. His small, open feet disappeared into the grass. When they emerged, closed, they were not empty.

Kenn didn't see the bird return to his perch. He was trying to decide whether to get out of his car and see what was going on. First he thought he would, then he thought he wouldn't. Then he thought he should.

After all, he didn't have anything else to do.

He opened the door, stepped from the car, and stared ahead. Sure enough, about two miles ahead there was a whole array of flashing red lights and what looked to be a fire truck.

"Why me?" he said, turning, looking at the very long line of cars behind him. "Why is it always me?" he pleaded, forgetting the hundreds of other drivers caught in the same predicament.

He did not see how the bird reached down and decapitated the vole and swallowed the head whole. He did not see how he left the perch and flew over the stalled line of traffic for a rendezvous with hungry young.

He was oblivious to everything but his frustration, which was real, and to his predicament, which was also real. As for the bird that had come and was now gone, it had no bearing on his life.

WHITE-TAILED KITE

TWENTY-TWO

Falco sparverius
AMERICAN KESTREL

A HOUSE SPARROW CHIRPED a warning. A Purple Martin vented its anger upon the incoming falcon with a snarl. And from the dark nook where the roof of the church improperly met the ledge, two heads emerged that were more than half hidden behind open mouths.

"*Eh-h-h, eh-h-h, eh-h-h,*" the siblings screamed in chorus. "*Eh-h-h, eh-h-h, eh-h-h.*"

Shoulder to shoulder, the young falcons crowded through the opening as their mother settled upon the ledge. But it was the older of the pair, the female, who reached Mom first and had a de-winged dragonfly stuffed down her throat as a reward. The loser, the male, got nothing, and his food-begging cries continued even after the adult had disappeared from sight.

In the church, the Reverend John Finch reacted to the noisy disruption and the swiveling heads of his congregation by raising his voice, finishing his sermon on the importance of traditional family values at levels approaching a shout. The Reverend's sermon, which included a discussion of the risks and hardships of single-mother parenting, would have interested the young kestrels. They were the products of such a family, and insofar as they were the survivors of what had been a clutch of four, they were also very lucky.

The unfed bird continued to cry and his sister joined in. True, she had just been fed. But a single dragonfly isn't lavish fare for a growing American Kestrel. It had been over an hour since she had last been fed and that meal had been nothing but a small grasshopper, no more than a mouthful. Her brother had been shut out of that feeding, too.

It is not common for nestling birds of prey to be so vocal. Hunger is a hard reality, but there are far greater dangers in the world, and noise attracts them. Even though the birds were twenty days old and each weighed as much as an adult, their lives were still at risk — from the minister's roaming cat or one of the pair of Cooper's Hawks nesting in the nearby woodlot.

As if this thought had suddenly occurred to them as well, the nestlings turned and retreated beneath the safety of the church roof. They took positions near the entrance of the man-made (but weather-enhanced) cranny, the male closer than the female. The next time Mom arrived with the goods, he did not intend to be runner-up.

Most young kestrels do not have the advantage (and associated risks) of an open ledge to run out upon. But nearly all enjoy the safety of a nesting cavity, and this explains their noisiness. Cavity nesters are safer from harm than are birds reared in open nests — even open nests that are well attended. If a hunting Red-tailed Hawk or scavenging crow does find a cavity filled with young kestrels, there is little it can do about it so long as the young remain within.

Natural tree cavities or abandoned flicker holes are preferred sites over much of the bird's range — a range that covers all of North America south of the tree line. But the birds are quick to use man-made structures (including nest boxes) where these are found in proximity to prey and open parcels of land. Holes in cliffs and even kingfisher nest sites excavated in streamside banks are also used.

Nesting in a protected fortress offers an advantage to adults, too. It means that they can spend more time hunting and less time guarding the nest, and hunt they must. American Kestrels raise big broods

157

"*Shoulder to shoulder, the young falcons crowded through the opening as their mother settled upon the ledge.*"

The Wind Masters

— three to seven birds per nest as opposed to two to four birds per nest for most raptors. Kestrels are also incredibly successful at parenting. Nine out of every ten birds hatched fledge. Small wonder the "killy hawk" is the most abundant raptor in North America.

Of course, when one member of a pair is lost, success goes down.

His mother was not in sight, but the young male started making food-begging calls again. He didn't call as loudly as before. The cry was closer to a single-note plea than a trisyllabic whine, but if anything it was more plaintive. Hunger had been the bird's companion since hatching, and during the first days out of the egg it had come close to mastering the bird's life.

The adult male bird had disappeared during the 28th day of incubation. He had gone to roost the evening before the first of the four eggs had begun to pip and never returned. Adulthood does not confer immunity from danger. The female waited until late in the afternoon before leaving the church to hunt for herself and her new daughter. She hunted the next morning also, returning with a field mouse for the cotton ball that was her daughter, and two sons still wet from the egg, exhausted by their struggle, and hungry. The fourth egg never hatched.

The first days of the family's life together were the most trying. It is the male who assumes the role of principal provider during the first week of a chick's development. The female's hunting is auxiliary; her main role is rendering prey into chick-size pieces and keeping nestlings warm. Like single parents everywhere, she found that alone she could not do all that was required, or at least she could not do it as well as two.

Even the weather conspired against the family and the female's best efforts. A late-season cold front moved into the area and stalled, bringing two days of rain. Hunting was poor. A pair of kestrels may make forty trips to a nest with prey during the course of a day. The female, hampered by conditions and with skills eroded by a month of doing little but incubating eggs, managed three per day, and the prey she carried — several earthworms, a beetle, and a tiny garter snake — could not begin to meet the food needs of her young. And then there was the cold. . . .

The unfed male stopped food-begging, stood, and retreated deeper

AMERICAN KESTREL

into the nest box. There are other biological demands made upon nestlings besides hunger, and he was feeling one of them now. Walking past his sister, and past the unhatched egg that had been pushed to one side, he leaned forward, defecating onto the cavity walls.

It was at this inopportune moment that his mother unexpectedly returned, carrying another dragonfly. Neither his dash to the door nor his vocal cries made any difference. He reached the ledge just in time to see his mother's departure and his sister draw her head back and swallow.

Among birds of prey, older, larger nestlings receive more food than their siblings, and the young female held both of these advantages, albeit by a slim margin. In American Kestrels, the size difference between males and females is apparent but not as pronounced as it is in many other raptors.

Size, however, is not necessary to determine the sex of kestrels, whether nestlings or adults. The plumage of nestling kestrels is different between the sexes; in fact, it is almost identical to the plumage worn by the colorfully patterned adults.

There are other North American birds of prey that are sexually plumage dimorphic — that is to say, males and females have markedly different plumages. These include the Northern Harrier, Hookbilled Kite, Snail Kite, and Merlin. But these differences are apparent in adult plumage; nestlings of these species look alike.

Of all North American raptors, only the sexes of juvenile kestrels wear different dress. And although easy recognition of sexes poses an advantage to birds seeking mates, what advantage this might pose to the nestlings of a cavity-nesting raptor is unknown.

On the ledge, in the sunlight, the differences between the two birds were very apparent even though the crown and scapulars were still coated with a frosting of down. The facial patterns were similar: rufous capped and double-moustache striped. The backs, too, were more alike than not — brown and barred.

But the flight feathers, almost fully developed now, were very different — brown and barred like the back on the female; gunmetal blue on the male. The tails, though hardly two inches long, were also different — characteristically brown and barred on the

The Wind Masters

female; rufous with a blue-black terminal border on the male. Both birds were handsome, but the male was stunning.

And hungry! — although the early privations suffered by the family had eased after the nestlings were a week old. In part, the situation was helped by the death of their sibling through starvation and exposure. It left one less mouth to feed and, as a matter of cold fact, the body was used to feed its two siblings. Raptors take a more pragmatic view of cannibalism — possibly because they are ignorant of Western cultural tradition; more likely because starvation is a hurdle they must vault now and again.

But the turnaround of the family's fortunes was also a matter of maturation. After one week, young kestrels can dismember small prey and feed themselves, leaving the female more time to hunt. Even in families where both parents hunt, by the time the chicks are half grown, females do much of the hunting.

The young female returned to the nest cavity, but the male remained on the ledge — hungry and frustrated. He picked up a discarded feather and dropped it. He picked at a rusty nail protruding from the overhanging shingles, but could coax little amusement from it. He stepped farther onto the ledge, keeping a watchful eye on the sky, turning his head sideways now and again to study a high-flying Purple Martin above or one of the homeward-bound churchgoers below.

The church, in the manner of New England churches, was on a hill, and from the ledge the young kestrel could take in most of the town, the river, and the fields and woodlands beyond. In a week, his wings would be strong enough to carry him to the places only his eyes could take him now. Then, for about three weeks, he and his sister would remain together, polishing the skills they would need for life on their own — and harrying their mother for food.

Absently, the bird started his food-begging cry again, softly at first, more loudly by degrees. His cries went unanswered and as his frustration increased so did his animation. He ran to the edge of the ledge, retreated, and then returned. He walked along its edge to its terminus and ran back.

He didn't see the crow that landed on the steeple of the church —

AMERICAN KESTREL

and the crow couldn't see him, although the crow knew where the nestling was and even what he was. The crow, after all, was also a resident. Crows know a great deal about the business of being crows, and it is the business of crows to know who is nesting where in town and how vulnerable they are. The crow also knew that the adult male was lost, the female had gone hunting, and the young were unguarded.

Having no better option, the crow set its wings and dropped, hoping to catch the young kestrel by surprise, which it did — but the surprise was mutual.

The kestrel was a good deal larger than the crow had anticipated and not as vulnerable as it had hoped; the overhanging roof saw to that. But the crow's greatest surprise was the discovery that the bird seemed fully alerted to the assault. A hawk's face was turned its way, and two black eyes bored into it. The crow hesitated.

Even a professional crow can be fooled. The face was not a face. The bill and the eyes were etched in feather, a head pattern that mimics a hawk's face, a ruse that sometimes daunts a predator when a kestrel's back is turned. The kestrel reacted to his danger a moment before the crow saw its error, but a moment was enough. This time, when it really counted, the kestrel won his race to the door.

The crow did not linger and the fright suffered by the kestrel did not last long. When the adult female finally returned to the nest with food, he was ready. His hunger and frustration more than matched the advantage of size enjoyed by his sister, and his reward, for screaming the loudest and getting his shoulders through the opening first, was the House Sparrow his mother dropped at the door.

A House Sparrow offers a substantial meal — more than enough to keep two growing young falcons occupied for a time. The adult kestrel took a perch atop the church steeple, the one once favored by her mate, and started to preen. There had not been many opportunities to rest during the last three weeks and she took advantage of this one. No telling when she might get another.

TWENTY-THREE

Falco mexicanus

PRAIRIE FALCON

JOE BOB, THE YOUNGEST, was loafing on his pa's favorite perch, trying to decide whether to chase the White-throated Swifts around, knock his brother off the nest ledge, or just sit out the New Mexico heat and digest his meadowlark quietly. His inclinations were running strongly in favor of dusting his brother.

Brother Sam Rayburn Mexicanus, who was the perennial object of Joe Bob's scampish humor (and who had not gotten any of the meadowlark), was sitting at the mouth of the sandstone cave where they, the most recent brood in a long line of Prairie Falcon nestlings, had been reared. He was toying with the idea of giving Joe Bob a little heat himself — but that shouldn't be too surprising. After all, he was a young Prairie Falcon. That is about the most hell-raising thing that a bird can ever hope to be.

Crystal, the female, was perched on a sandstone spire high above her siblings. She was tending to her new flight feathers, turning her large, dark falcon eyes upon the horizon, giving serious thought to seeing what was over there.

163

After all, she was a big girl, even bigger than her mother. She was able to fly (as well as fledgling Prairie Falcons *can* fly), able to chase things, almost able to hunt for herself . . . if she had to (which up to now she had not). Looking down at her great, taloned feet she seemed almost surprised to see the remains of the thirteen-lined ground squirrel that her mother had carried in around midmorning.

"When *I* start hunting," she more or less thought, "*I'll never* hunt these old ground squirrel things that Ma keeps bringing."

"When *I* start hunting," she more or less promised, "I'll only kill meadowlarks and doves like Pa."

Her aspirations notwithstanding, chances were that Crystal would be eating a good many ground squirrels in the course of her life. Prairie Falcons are versatile hunters, able to take a variety of avian and mammalian prey. But they are practical, too. These "desert falcons" of the American West capitalize on whatever happens to be the most available. Ground squirrels, when their populations flourish, are the falcon's primary prey. Open-country birds, *particularly* flocking birds such as starlings and Horned Larks, are popular winter prey items.

The techniques Prairie Falcons employ to capture prey are likewise varied. They can hunt from perches or cruise the contours of the land, surprising prey and securing it in a swift, one-sided dash. But the Prairie Falcon's trademark technique, the one that weds it to the open, coverless western lands, is the stoop and stealth glide.

The sand-colored falcons tower up over the prairie, the desert, sagebrush flats, or pleated agricultural lands. Set against a cloudless sky, where depth loses all bearing and heat waves dissolve forms, the birds fairly disappear. Only the dark underwing coverts, the classic field mark of a Prairie Falcon in flight, may be visible. Once prey is sighted, the falcons fold and stoop — but obliquely, not directly upon their prey.

That would be too obvious. That would not defeat the eyes of prey or suborn the open honesty of the land. The falcon stoops short, falling to earth several hundred yards shy of its mark, then levels out, just above the ground, closing the final distance in a set-winged, eye-defeating glide. The sand-colored bird and the sand-colored land

The Wind Masters

become one. Often enough, a ground squirrel has the first inkling that all is not right with the world the moment it leaves it — double-wrapped in raptor talons.

Releasing her grip, Crystal let what was left of the squirrel (which despite her professed distaste wasn't much) fall. It dropped past Sam Rayburn, who launched himself to intercept it, and Joe Bob, who had pretty much decided to harass his brother anyway.

At this point Crystal realized that even though she didn't necessarily want the old squirrel herself, she wanted her brothers to have it even less. After all, the squirrel *was* hers. The ensuing melee, more an aerial brawl than a sibling test of reflex and coordination, covered both sides of the canyon and lasted ten minutes.

Although adult and juvenile Prairie Falcons are similarly plumaged, differing not in pattern but in shades, the clumsiness of the trio marked them for what they were — young falcons, little more than a week on the wing. Their flight was bold, even reckless, but it was not breathtaking. Their turns were sweeping, not tight, and their stoops were somewhat tentative, as if even now they weren't quite sure how to stop or what to do should they actually intercept their target.

They were just having fun, as young creatures will — serious, skill-honing fun. They would need these skills in a few short days when they would leave the nest site and test themselves against the world. All three birds were still in the air, looking for the next diversion, when Ted arrived on the scene.

Ted was a young Peregrine Falcon, an *anatum* Peregrine, one of the brood from an old traditional nest site a mile up the canyon. Only several days off the ledge, this was his first day of exploration.

The Peregrine's nest site was superior in many ways to the location used by the Prairie Falcons — higher, drier, cooler, better protected from the ravages of afternoon thunderstorms. The cave used by the Prairie Falcons was set on the south side of the canyon, to avoid the direct rays of the sun. Because the lower part of the canyon heated so quickly, the positioning and recessed confines served to reduce heat stress upon the young.

During the DDT era, when western Peregrine numbers fell to a vestige, several pairs of Prairie Falcons did, off and on, occupy the

"Their flight was bold, even reckless, but it was not breathtaking."

The Wind Masters

old Peregrine aerie, raising successive broods there. But when Peregrine numbers rebounded and a prospecting pair of Peregrines reoccupied the site, the Prairie Falcons moved out.

There were other potential sites in the main canyon that would have served the evicted falcons' needs. That wasn't the problem. There was even enough prey in the adjacent desert — another factor that limits the nesting density of raptors in an area. That wasn't a problem either.

The problem was the canyon itself. It was small, straight, and narrow — no more than a quarter-mile wide at the mouth; less than two hundred yards wide where the Prairie Falcons nested. Up-canyon birds going in and out were forced to pass through the nesting territory of down-canyon pairs, and whereas birds of prey may tolerate another of their species or a very similar species in their larger, hunting territory, tolerance in the vicinity of the nest is another matter. Two pairs of falcons was the canyon's amicable limit and how amicable was debatable.

The down-canyon Prairie Falcons and the up-canyon Peregrines feuded more or less constantly. While the Prairies had little call to travel up into the canyon, the Peregrines were obliged to pass through Prairie Falcon airspace as a matter of routine. This was the source of friction between the factions. Altercations were commonplace and intense.

Even while incubating, the female Prairie Falcon would leave her clutch of five eggs and follow the later-nesting Peregrines back to their ledge — screaming all the way. When the birds neared the Peregrines' aerie, the up-canyon birds' territoriality would surge and they would turn on their pursuer. More than once the birds had grappled and over the course of several nesting seasons had drawn blood.

Though the birds were fairly evenly matched, the Peregrines had slightly more finesse. They were masters of the dance and jab, boxers in the European style. The Prairie Falcon wasn't exactly a street brawler (although Prairie Falcons do tend to bind to their prey as opposed to striking it in the manner of a Peregrine). She was also as territorial as the Peregrines and as temperamental. But at the end of each aerial bout, when all the points were tallied, the Prairie Falcon commonly took more than she gave. Despite her best efforts, she never succeeded in getting the Peregrines to do much more than hug the far side of the canyon when they passed.

Whenever the female Prairie returned to her cave following a diplomatic harangue with her up-canyon neighbors, the other denizens of the canyon walked on eggshells. A Prairie Falcon is a moody, flash-tempered creature under the best of circumstances, but an aroused Prairie Falcon is retribution on wings. The canyon's Red-tailed Hawks soon learned to read her mood from afar and steer clear. The young Barn Owls, not a hundred feet from the falcons' cave, pressed against the back wall of their grotto.

Only the agile American Kestrel, who nested in a sycamore directly below the falcon aerie, seemed indifferent to the female falcon's wrath. In fact, when the female Prairie Falcon was in a snit, it seemed to fire the ire of the kestrels, who would harass her until she returned to the cave.

It's difficult to say what the young Prairie Falcons thought about Ted — whether they had inherited something of their mother's antipathy for intruding falcons, or whether Ted represented nothing more than a fresh new sparring partner. Whatever their regard, the Prairie Falcon kids stopped feuding, joined ranks, and went for their up-canyon neighbor like toughs meeting the new kid on the block.

Ted had siblings of his own and was not a total stranger to the sparring advances of young falcons. But he had never in his few short days on the wing faced anything like *this*. The Prairie Falcon kids swarmed all over him. They were above him, below him; charging head on and swiping at him from behind. He'd rush for one and have to fend off two. He'd swerve to avoid a collision with Joe Bob and run smack into Crystal.

It cost a few feathers, it cost pride, but most of all it cost Ted altitude. Every time the young Peregrine was forced to avoid a collision or obliged to turn on his back to ward off an attack, it brought him closer to the ground. Confused, tired, and out-boxed three to one, Ted had little choice but to land on the boulder-strewn canyon floor.

This, to Ted's mind (and the chivalrous code of air combat), should have ended the matter. But it didn't. Where Peregrines hold grounded targets in contempt, Prairie Falcons consider them prime fare. Where the Peregrine considered his grounding the falcon equivalent of raising a white flag, the Prairie Falcons considered it new rules of engagement.

The scrappy trio continued to throw combination punches at Ted's head, forcing the increasingly befuddled Peregrine to duck repeatedly. Finally, as an act of sheer unhappy desperation, Ted was forced to take shelter between two boulders.

The indignity was terrible, but the maneuver was successful. It took much of the fun out of the Prairie Falcons' game, and, as with kids everywhere, effort without fun becomes not worth doing. When Ma Prairie Falcon arrived on the scene with prey ("Yes," Crystal noted unhappily, "another ground squirrel . . ."), even the irascible Joe Bob was ready for new diversions.

The female Prairie Falcon circled over the creek, waiting for her young to climb into position. Before the hungry horde could reach her, she let the squirrel fall. Overanxious Sam Rayburn missed. Crystal made a halfhearted grab and lost it. Joe Bob nailed the rodent with a neat, one-talon grab and then headed for the cliff, his siblings in pursuit.

The female Prairie Falcon did not leave, as she might have done, nor did she perch, which she commonly did after a successful hunt. Instead she folded her wings and stooped toward the creek bottom, cacking angrily. She stooped on the huddled and befuddled falcon several times but was no more successful at dislodging the intruder than her young had been. She flew up and down the creek bed several times, searching for something to vent her anger upon. Finding nothing, she flew to the boulder-strewn sandbar and landed on a water-sculpted stone.

The adult Prairie Falcon and the young Peregrine regarded each other. Neither knew what to do. A student of raptors would have noted many similarities between the two species. The large, dark protruding eyes and the prominent superciliary ridge above them; the spiked nostrils and the tomial teeth — the things that make falcons falcons. But for the moment the two falcons could see nothing but the things that distinguished them. The blue-gray back and helmeted head of the Peregrine; the sandy brown back and sideburn-slashed face of the Prairie. The larger size of the female falcon and the smaller, more compact form of the young male. Both the similarities and the differences combined to drive a wedge between them.

But there was one difference that was as binding as the others were divisive. It drew the two birds together as strongly as territoriality drives two species apart, and as time passed, anxiety with it, this difference grew to dominate the birds' regard for each other and set other differences aside. One bird was a parent; the other was a youngster who was used to the tending care of adults. For weeks, this biological relationship had dominated their respective lives. All that stood between the two falcons was ten feet and the chasm that lies between two species — however wide or narrow that might be.

The young Peregrine moved first — or maybe he reacted to some subtle sign of acquiescence on the part of the Prairie Falcon. It would be difficult to say. Suddenly, the young bird lowered his head in a manner of supplication or appeasement. He left the protection of the boulders, mincing forward, and began making food-begging cries. Both his manner and his cries were not unlike those of young Prairie Falcons wanting to be fed.

The female Prairie Falcon was somewhat taken aback, but she did not attack the youngster — in fact, when the bird approached so close that he crowded her, she was forced to retreat to one side. When the young Peregrine followed, she flew, leaving the hungry Peregrine behind.

Tolerating another family of falcons in the valley was one thing. Tending their young was another.

The young Peregrine stopped his food begging as soon as the Prairie departed. Then, remembering his wings, he lifted off the sandbar, circled once, and headed up-canyon. There had been enough adventure for one day.

Joe Bob would have chased him but he was busy with the squirrel. Sam Rayburn might have chased him but he was busy trying to get part of the squirrel.

Crystal wasn't interested in chasing at all. She was staring at the horizon that she would be visiting any day now. The canyon was growing too small.

Buteogallus anthracinus
COMMON BLACK-HAWK

IF YOU CONCENTRATED upon the smell of water, the coolness of the air trapped beneath the canopy, the sound of wind through the willows, and the mad mutterings of streamside Yellow-breasted Chats, you might guess that the riparian woodland was in Wyoming, Oklahoma, or the Central Coast of California. But to draw any of these conclusions, you would have to ignore the flinty air and the pastel-colored canyon walls reaching high above the cottonwoods, pinching a moisture-starved sky. You would also have to disregard the querulous whistles of Phainopeplas and turn a blind eye to the glowing garb of Vermilion Flycatchers.

Most of all, you would have to ignore the imposing raptor perched on the branch of a sycamore overhanging the stream, because to find this bird in the United States is to narrow your geographic options to a select few locations: Jeff Davis County, Texas; a wedge of western New Mexico; and a handful of riparian corridors in Arizona. The bird is a Common Black-Hawk. There are fewer than 250 breeding pairs north of Mexico. All but a handful are to be found along the

Big Sandy, Virgin, and Gila river drainages and along streams that lie in central Arizona's Mogollon Rim.

Despite its size — which is large, Red-tailed Hawk size — and its color — which is what some other hawks merely appear to be, black — it would not be difficult to overlook the bird. Common Black-Hawks are sedentary creatures. Perched, on streamside boulders or limbs overhanging streams, is their natural state — except during courtship. In courtship, hormones boost the birds aloft, where pairs stall, dive, and flutter, displaying bright yellow legs and a golden cere to their best attractive advantage.

But courtship was the furthest thing from the bird's mind now. It was August, rest time in the natural cycle for Black-Hawks. The skyrocket displays of March were past and their memory dulled by the demands of parenting that were mercifully over — well, *almost* over.

From this favorite perch, on the U-shaped branch of a cottonwood straddling a rippled section of stream, the adult, a female, could see one of the pair of young she and her mate had raised. The youngster was perched downstream, where he could watch both the adult and the stream. Both, to his mind, were sources of food.

As she watched, the young bird grew suddenly attentive, then animate. First he turned his head, studying the stream below. Then he leaned forward, poised like a runner at the starting block. Then he righted himself, sidling along the branch that served as his perch. Then the bird dropped, more than stooped, toward the water below.

Either the young bird misjudged the angle (which is likely), or his intended victim was faster than the reach of his talons (which is also likely), but the bird missed. More at a loss as to how to proceed than surprised, the youngster stood, for a time, belly-deep in water, wings splayed to the side. After several moments of contemplation he stalked along the stream course, hoping to relocate the small fish, the Gila sucker, that had eluded him. Failing this, he hoisted himself onto a gravel bar, wings drooped, to air-dry and to consider life's most recent failure.

An observer unfamiliar with Common Black-Hawks and seeing the two birds might conclude that the adult and the immature were different species. Certainly, insofar as plumage went, the two had little in common.

The Wind Masters

"*This favorite perch, the U-shaped branch of a cottonwood straddling a rippled section of stream . . . .*"

173

COMMON BLACK-HAWK

Where the adult was uniformly charcoal colored, the youngster was brown above, buff below, and richly spotted. The face was pale, accented with a bold brown line through the eye and a mutton-chop stripe. The dark-tipped tail, more patterned than banded, was not a bit like the black tail with the broad white band typical of adults. While it is not uncommon for immature raptors to have plumage characteristics that distinguish them from one or both adults, these differences are not usually as pronounced as they are in this tropical buteo species.

As dramatic as the plumage differences were, they were nothing compared with the gulf separating the skill levels of the two birds.

Even though the young bird had inherited all the tools necessary to meet the challenges of life — agility, keen eyesight, strong talons, and a tearing bill.

Even though he occupied prime foraging habitat, one filled with a veritable shopping list of prey species.

Even though these dietary items were not as agile as the prey species consumed by many other raptors, the youngster's hunting skills were insufficient to meet his needs. An ability to hunt may be innate in birds of prey, but the techniques birds employ to capture prey must be learned and honed.

The young Black-Hawk's older sibling, a female, had departed several days before, following the watercourse west. Her skills were hardly more developed than his, but she was more opportunistic and less aquatic in her focus. The stream was up, flooded by seasonal monsoon rains. This made finding and catching such Black-Hawk dietary mainstays as fish, crayfish, and aquatic insects difficult. It made preying upon things such as beetles, lizards, centipedes, and recently fledged nestlings a more productive prospect. It made it possible for the female to break her dependency upon the adults and wander — a thing young raptors, like young humans, eventually do.

The adult, who would not wander and would not leave her territory until it was time to migrate in October, was not daunted by the high water in the stream and was not interested in beetles or centipedes — although a lizard would tempt her or a frog or a snake, so long as it wasn't venomous. Though hardly less opportunistic than her wandering youngster, she could afford to be more particular about her diet because unlike her progeny she was a *very* skilled hunter, and she had a particular talent for catching fish.

The first principle in Black-Hawk technique is *immobility*, stealth. Other birds of prey use speed or surprise as their first resort, but Common Black-Hawks are perch-firsters. Only when prey is sighted do . . .

The leopard frog, who until that time had lain motionless and unseen, leapt into the stream and kicked its way to the edge of the eddy. It stopped at the point where the current grew swift, floating on the surface, trying to decide what to do.

It wasn't the fish she wanted, but it was prey, and the adult left her perch in a silent, close-winged glide that carried her to a branch of the log that was the source of the eddy. This was the application of the second law of Common Black-Hawk hunting technique: when prey is sighted, reduce the distance.

The frog dove immediately, leaving a cloud of sediment hanging in the water where it retreated. And although it was out of sight, and although this subterfuge might have worked with a less experienced predator, unfortunately for the frog it was not out of mind and it was not out of the reach of the bird's bare tarsi. The shadow that passed over the submerged amphibian and the talons that closed around it were one. The frog was dead before the bird reached her sentinel perch.

If there were justice in the world, which may or may not be the case, and if it applied to Common Black-Hawks, which doesn't appear to be the case, the bird would have been allowed to enjoy the fruits of her success unmolested. But the bird had barely retaken her perch when her youngster was beside her, mouth open, head down, making a piteous racket. He would have done this whether he was hungry or not, but the fact was that the bird was hungry. It comes of being inexperienced.

The adult, who was also hungry, hesitated, caught, as Immanuel Kant might have phrased it, in a conflict of maxims. She wanted the frog for herself. But she was still a parent susceptible to the demands of her young. For some species of hawks, principally nonmigratory raptors, the exchange of food between parents and young may even continue until the next breeding season.

In the end, the adult's decision was a lopsided compromise. She dropped the frog, letting it fall to the stream below. Then she left. The young bird didn't quite know what to make of this departure from common practice, didn't know whether to follow the adult who had always been the source of food or the food itself. In the end, he did the right thing. He went for the frog.

The adult flew along the river on practiced wings. In flight, the broad white band on the tail was almost eclipsed by the astonishing breadth of the bird's wings.

She and her mate had defended this stretch of river for five summers, and she knew every nuance, every perch and curve. It wasn't a large territory, several hundred yards of river, no more, and it is typical of Common Black-Hawks to maintain relatively compact territories. Maybe this is linked to the small clutch size of this riparian buteo — between one and two eggs per nest; maybe it relates to a dearth of suitable habitat. The two are not necessarily unrelated.

The bird took a perch on the eastern end of the territory, one favored by her mate — who also was partial to fish but who, of late, had developed a taste for crayfish (which were more abundant on the western end of the pair's range). She waited several moments, waited until the family of Cassin's Kingbirds wearied of their scolding, then setting her wings, she executed a long, straight-winged glide to a sandbar near a quiet pool and lit a dozen feet from the river's edge. In the river there was a momentary swirl, then nothing.

It had taken the bird years to learn the art of fishing. Other birds, principally herons and storks, are practiced masters of the art, but raptors adept at the time-honored tradition of luring fish are few. Those birds that do count fish in their diet, such as eagles and Ospreys, are more hunters in an aquatic medium than fishermen.

Very cautiously, very deliberately, the bird approached the river until she stood poised at the water's edge. Very slowly, she extended one wing until the tip touched the water. Around the wing, sunlight rippled and her efforts to pierce the surface with her gaze were deflected by shards of light. Beneath her wing, the water lay in shadow and her eyes plumbed the bottom.

In the sun-heated water, the shade alone would have been sufficient to attract fish. The hawk had discovered this penchant of fish to seek

The Wind Masters

shade by accident, as a young bird, back in the days when overzealous efforts to secure prey often led to wings that needed drying. But in time the bird had added an ingenious refinement to the fisherman's art. One that had come about through a combination of accident and perceived opportunity; one that comes very close to something humans call *cause and effect.*

Abruptly, the bird began waving the tip of her wing in the water, troubling the surface, and around the wing, rippled rings spread outward as if some small creature were struggling on the surface. From several quarters, the shadows of fish drew close to investigate.

The bird bided her time. Patience is a Black-Hawk virtue. She waited until several fish had moved beneath the wing and edged to the surface. Then she thrust out a net of talons, closing a fish in her grasp. This time when the young bird appeared, with his crop still bulging from the frog he had consumed, he was given no more than the opportunity to watch while the adult bird fed.

Pandion haliaetus
OSPREY

REARING LIKE SOME BROWN AND WHITE COBRA of the air, the hawk checked her circling flight and started to hover. Yellow eyes fixed, tapered wings warding off gravity, she studied the surface of Block Island Sound and the phantom forms below.

It takes energy to hover in place. But it takes more energy to hover in place, attempt to capture prey, and fail. So the hawk was taking her time. Making the careful calculations that were necessary to secure her very specialized prey. The analysis might have read something like this:

DISTANCE TO SURFACE: 70 feet (an average hovering height for Ospreys)
DEPTH TO TARGET: 2.5 feet (a little deep — but doable)
ANALYSIS OF TARGET: Menhaden class, medium-size fish; approximate weight, one-half pound (well within tolerance specs)

TARGET'S CRUISING SPEED: 3 miles per hour
TARGET'S HEADING: 290 degrees
CONTINGENCIES: None

The bird did not need to make allowances for refraction — the light-bending sleight of physics that causes an image to appear where it is not. The bird was an Osprey, a fish hawk! The bird knew all about refraction.

Very suddenly, very correctly, the bird drew her wings back and her head down. Embracing gravity, she plunged. Though her wings were not fully closed, the long tips trailed well behind the tail.

The angle of the dive was steep, almost 60 degrees, but Ospreys can execute near vertical dives if necessary. The bird's speed as she neared the surface was almost thirty miles an hour (but dives approaching fifty miles per hour have been clocked). To the untrained eye it seemed that the dive was seamless — an untempered line drawn between predator and prey. But the truth was that the Osprey made a number of mid-dive microadjustments — fitting the dive to the target until they were a match.

Just before impact, the bird's feet lanced forward — twin sets of talons that projected ahead of the bird. They absorbed the first shock of impact and continued down. All the weight of the bird, some 1,600 grams, fell in behind them and continued beneath the chop.

The splash was still settling when the bird's wings broke the surface and pushed downward, catapulting the hawk back into the sky. Except for the beaded strand of water droplets that slipped from her balled feet, the talons were empty.

Several hundred yards away, another Osprey, an adult, replicated the movements of the younger bird but with more finesse and one important difference. When the adult emerged from the water of the Sound, it held a ten-inch menhaden in its talons — a consequence that was not unexpected. The hunting success for an adult Osprey is about 75 percent.

It was easy to tell the adult Osprey from the immature, although plumage-wise, adults and immatures look much alike. Both birds were charcoal brown above, white headed and white bodied below.

"*Except for the beaded strand of water droplets that slipped from her balled feet, the talons were empty.*"

The Wind Masters

Both brandished sparse, dark, shaggy crests and a prominent, horizontal racing stripe drawn across the eye.

And both were large, almost eagle sized — although Ospreys are not "eagles" any more than they are "hawks." In the taxonomic hierarchy, the Osprey is placed in its own distinct family, Pandionidae — the only North American bird of prey so distinguished.

More than plumage and size, more than the pale feather edges on the back of the younger bird or the more prominent pale band on the tip of its tail, the thing that distinguished the adult Osprey and the immature was the fish. The adult, an accomplished fisherman, had it. The juvenile, a mere three weeks out of the nest, did not. Although the young bird was endowed with all the innate ability and anatomical refinements she needed to be one of the planet's most specialized birds of prey, she still lacked one important quality. Skill. The ability to make use of the advantages she had — which in the case of the young Osprey were considerable. These included:

Long, gull-like wings, contoured for energy-efficient flight and strong enough to withstand the impact of an Osprey's dive.

Talons that were long, deeply curved, and sharp as the spines of the fish they impaled. They projected from toes that were grotesquely falcon sized, but stronger and more versatile. The outer toe on either foot could be folded back to cast a wider net at prey, and the pad of each foot was studded with tiny impaling spines — an anatomical refinement well suited for capturing slippery prey.

Each murderous foot was wielded by an exceedingly long leg, giving the Osprey a boardinghouse reach to complement dives that sometimes propelled the bird three feet beneath the surface. No other birds of prey, not even those that count fish among their diet, are capable of such a plunge. A Black-Hawk or an eagle attempting to duplicate the feat would pay, at least, a sodden price.

Ospreys, on the other hand, are immersion-safe; the fish hawk's feathers are oily and dense. Repeated immersion will defeat these aquatic coveralls. Heavy rains will breach the covering's protective integrity. But an Osprey emerging from an everyday dive sheds water like an eider.

These were the evolutionary advantages that the Osprey could bring to bear. Not only did they confer upon the Osprey the ability to secure prey beyond the competitive reach of other birds of prey, they made it possible for Ospreys to colonize much of the world. Europe, Asia, Africa, North America, Australia — all the continents but Antarctica and, curiously, South America (where North American Ospreys spend the austral summer). Among birds of prey, only the Peregrine Falcon (another prey specialist with a worldwide resource to exploit) is more cosmopolitan.

In North America, the Osprey is widely distributed from Atlantic Canada to western Alaska. In the United States, breeding birds are more concentrated along the Atlantic and Gulf coasts, the Pacific Northwest, and western portions of the Great Lakes. Wherever they reside, Ospreys are never far from water — indeed, it has been aptly said that Ospreys require only three things to breed:

Ice-free water for the duration of the breeding season. Slow-moving fish that feed close to the surface. A nest site that is relatively molestation free.

Most Ospreys find the sanctuary they need for nesting in trees; denuded or top-sheared spires that allow an unencumbered approach are preferred. But Ospreys are eclectic in their approach to nest sites, adopting all manner of natural and unnatural substrates — so long as they lie close to fish-bearing water. Ospreys are also relatively untroubled by humans, which is fortunate. Since humans, like Ospreys, also tend to concentrate along coasts, tolerance of humans in a fish-eating bird of prey is an advantage worth having.

Anatomical refinements and an open-minded attitude were not the only advantages that the young Osprey carried into the world. The bird enjoyed the privilege of a very special benefactor called "good fortune" by some and "luck" by others. The ten-week-old bird had benefited from assorted fortunate accidents that had helped her secure a foothold in the universe (even if her foot had yet to encircle prey).

It was fortunate, for instance, that the bird's colony, on Great Island, near the mouth of the Connecticut River, was well placed — buffered by water and marshland against the egg-snatching predilections of raccoons.

It was fortunate for the young Osprey that her parents were older

The Wind Masters

birds, both in the second decades of their lives. As older birds, whose experience as breeders was refined by the successes and failures of many seasons, their ability to raise young was superior to the efforts of less experienced pairs — and it showed. In the volume of food brought to the young Osprey and her two siblings; in the care the young birds received.

Even during periods of high winds or turbid water, conditions that hamper an Osprey's hunting success, the young Osprey and her two brothers were fed. Other chicks, in other nests, were not, with the result that many of the smaller, weaker siblings (usually the last birds hatched) starved.

When the bay was raked by a vicious thunderstorm and several tree nests toppled, spilling eggs and nestlings, the young female's nest weathered the storm. In part, this bit of fortune was related to the structure supporting the nest — a platform specially designed for this purpose by state biologists. But the nest itself was well constructed and well balanced. The wind could find no weakness to exploit.

It was also fortunate for the young Osprey that she had been born in an enlightened age, one in which birds of prey live relatively unpersecuted lives. In the past, Ospreys suffered assorted human persecutions — shooting, pole trapping, egg collecting — practices that are now deemed unacceptable by society and forbidden by law. The species' current, elevated esteem is rooted (ironically) in a chemical agent that almost resulted in the Osprey's demise: the substance DDT, the same chemical that had caused Peregrine Falcon numbers to plummet. Had the egg that encased the young female Osprey for 37 days been deposited before DDT was banned, chances are it would not have hatched. Nearly 20 percent thinner than a healthy egg, it would have shattered beneath the weight of the incubating adult. Between 1946 and 1972, productivity in the colony virtually ceased. Where 150 nests once flourished, only a handful remained.

The demise of the Great Island colony was not isolated. It was part of an epidemic. Across Europe and North America, Osprey numbers plummeted. The decline resulted in a tremendous public outcry, the ultimate banning of DDT in the United States, and the species' slow recovery. The umbrella of popular acclaim that came as a result of the near tragedy has endured.

There are, in the world as Ospreys know it, sad fortunes that good luck had deflected away from the hunting young female. Two nests in her colony had fallen victim to predation by a Great Horned Owl. One incubating adult was killed. In another nest, a three-week-old bird had become enmeshed in a tangle of fishing line carried in by adults. The bird did not strangle so much as starve.

Ospreys whose lives are not cut short by ill fortune are long-lived birds. Among breeding adults, annual mortality is less than 15 percent, and in the wild some individuals have been known to survive twenty-five years or more.

But among young birds death is more intimate, accounting for 50 percent of all birds in the first year of their lives. It takes two years for an Osprey to reach breeding age. For the majority of subadult birds, this period is spent in Central and South America, where most North American Ospreys spend the austral summer. Only birds nesting in Florida, along the Gulf Coast, and in the waters of Baja California are resident throughout the year.

From the time they leave the nest, young Ospreys must endure tremendous stress and surmount a battlefield of hardships. Starvation is a constant menace for inexperienced birds. The dangers and energetic demands of the species' long migration to and from Central and South America levy a harsh toll. Shooting outside of the United States is still all too prevalent, and the use of biocides, including DDT, is still widespread in Latin American countries. By the spring of a generation's third year, attrition has whittled the ranks. Only one out of every three nestlings usually survives to breed. All it takes to be counted among the lucky ones is innate ability, honed skills, and luck.

The young bird, whose ability was sound and whose luck had thus far been good, was still in the process of honing her skills — flying, hovering, diving, lifting off the water — all the skills an Osprey needs to execute one of life's most important endeavors: catching fish.

In the days since fledging, she and other young Ospreys had frequently foraged offshore, drawn by instinct and by the adults who

The Wind Masters

went out to intercept the schools of summer bait fish. Like the other wheeling, calling, and clumsy juveniles, she had flailed her wings and flop-splashed into the Sound — sometimes with, sometimes without a target in mind.

Once, she had become so waterlogged that she was forced to swim ashore. Twice, she had secured injured fish left on the surface in the wake of blitzing bluefish. She'd thought of these accomplishments as "captures" (but the truth was they were closer to scavenging).

Yet the day before, plunging blindly into a bluefish-pushed school of menhaden, the youngster had raked a fish — planted a talon but missed on the grasp. The near success was as much accident as design, but accident is sometimes the companion to success. It even has a name. Luck.

Encouraged, the young female tried again . . . and again. . . . By the time the school sought the safety of deep water, by the time the tired, sodden young Osprey lumbered back to land — and the adults who would feed her through much of August — she had learned much about technique. Though her talons had yet to close over anything more tangible than failure, by the time the last school of bait fish had dispersed, her failures had become as much a matter of accident as her earlier near success.

Climbing to sixty feet, holding into the wind, moving slowly, the young Osprey studied the water below her. Searching for a fish that met her criteria. Surface feeding. Slow moving. Less than one pound in weight.

Although Ospreys can and occasionally do capture larger fish, it is of dubious benefit. Ospreys can consume only about ten ounces at a sitting. Unless an adult bird is hunting for a nest of hungry young, more fish only means more work.

Several times, the young Osprey spotted the shadowy forms of fish. Checking her forward momentum, hovering to hold her place, or dropping lower to shorten the odds, she calculated an attack — but missed her opportunity as the fish moved beyond her reach.

Once, she caught sight of a host of shadows — shadows of size. She hovered to a halt. . . . Dropped. Stopped. Dropped. . . . But did not stoop.

185

A meter in her mind told her the fish were too large. Too large to secure. Too large to lift out of the water. Too large to carry. The bird weighed almost four pounds. The smallest of the shadows would have equaled this.

She did not stoop. But when the shadows lunged ahead, turning the surface of the bay to jagged madness, she was ready, hovering less than thirty feet above the melee.

Beneath the churning surface, crazed bluefish were cutting through a mass of menhaden. The silver-sided fish bulged to the surface and lunged to avoid the predatory school. Gulls gathered above the mass, snatching pieces of fish so freshly severed they didn't know they'd been cut off from life. Common Terns arrowed into the fracas, snatching small, panicked fish from the mouths of other fish.

The Osprey could find no target in the free-for-all — at least no target more specific than the churning mass of fish. She stopped hovering and plunged, feeling the bump of fish against her outstretched talons but coming away empty.

Luck. Bad luck.

Reaching up, then down with her wings, breaking the ocean's clutch, she climbed once more above the blitz, folded her wings, and fell. This time an open foot found living prey. She felt one talon slam home. She felt the electric thrill that shot from her foot to her brain and back to her foot and said, in a fraction of a second: CLOSE! But in the subfraction of a second before the order could be executed, she felt the fish slip free.

More bad luck.

She floated on the surface for a moment, caught between the flurry of birds and the frenzy below. It took the slightly waterlogged bird two pushing strokes to break the ocean's bond. As she climbed she roused . . . twice, sending water droplets flying.

Climbing to forty feet, hovering without circling, she positioned herself above the mass and prepared to try again. Beneath her, the water fluttered and then it roiled as fish, in their excitement, reached the surface. She turned and dove.

All of her innate skill went into that dive. All of her acquired skill. It was flawless in its design, flawless in its execution. A tribute to the

The Wind Masters

compounded fortunes that had brought her to this junction — the make-or-break point in any young raptor's life.

At the last possible second, as she threw her talons forward, the dark back of a fish took shape in the madness and she had just the time it took to micromanage her flight — and drive her splendid talons home. Home!

She felt the talons sink deep. She felt her grip lock. She knew even before her body slid beneath the waves that she had succeeded.

It was only bad luck, and not poor judgment, that had directed her talons into the back of a bluefish that weighed twice what she did. It was only the pendulum swing of fortune, and it was moving the other way.

TWENTY-SIX

Buteo albonotatus
ZONE-TAILED HAWK

THE SUN HAD VAULTED the canyon wall, but the clouds that are spawned by the higher mountain peaks had yet to form. Below the crest of a yucca-studded hillside, near the mouth of the canyon, a White-winged Dove landed near a pool of rainwater and started to drink.

Though young, she was alert, having learned in the weeks since fledging that the world is a perilous place. When the Prairie Falcon that nested on the nearby cliffs began turning circles overhead, she took momentary cover in a stand of ocotillo. When an Acorn Woodpecker sounded a warning, she froze . . . and watched as a Cooper's Hawk flew up-canyon with the remains of a Black-headed Grosbeak dangling from its talons.

"Whew," she felt more than thought.

Interruptions notwithstanding, the dove finally satisfied her thirst, and finding no danger imminent, she decided to take a moment and tend to her feathers. She remained alert, of course. But from her vantage there was nothing to be seen that hinted of danger — nothing but a small contingent of Turkey Vultures turning lazy circles against the peaks. The vultures lived among the limestone cliffs that guarded the mouth of the valley and they posed, she knew from experience, no threat to her.

As the vultures moved out onto the open plains, drifting behind the hillside and view, the dove turned her attention to her morning toilette — straightening a few barbules on a tail feather, tuning a couple of primaries that had been giving her trouble. Her working parts in order, she turned her attention to the iridescent feathers lining her breast.

This was her favorite part in the preening ritual, the part that gave her the most satisfaction, and she usually saved it for last.

"Ah, bliss," she felt more than thought.

Her bill was well buried and the nictitating membrane of her eyes half-closed when the sun was suddenly eclipsed by shadow.

"Sun must have gone behind a cloud," she assessed more than thought. In another moment the qualifying rejoinder "but there are no clouds" would have put her to instant flight.

But she didn't have that moment.

It wasn't lack of vigilance that betrayed the dove. She had, in unfortunate fact, seen the very shadow that was about to eclipse her life. It had circled with the vultures and she had dismissed it.

The agent of her demise was a Zone-tailed Hawk, a buteo and a bird-and-lizard specialist. This is a difficult line of work for any bird of prey but a particularly difficult specialization for a buteo.

Some bird-catching raptors rely upon speed to capture prey; some depend upon agility. All to a degree court a margin of surprise to help secure creatures that are often faster and more agile than they are. But in all of North America, only one raptor links surprise with disguise. This is the Zone-tailed Hawk of the American Southwest, which in appearance and mannerisms mimics the harmless Turkey Vulture. Many birds, and not a few reptiles and small mammals, fall for the ruse and then fall prey to this raptorial grifter.

If the dove had been granted the time to turn around and study the

"*The vultures posed, she knew from experience, no threat to her.*"

The Wind Masters

bird bearing down upon her, she would have confronted a broad set of talons projecting from a soot-colored bird of prey. This is hardly the sort of thing that inspires casual regard in a dove and hardly the image the bird of prey had presented when it circled with the vultures.

What the hawk had projected, and what the dove had seen, was a large, dark, long-winged, long-tailed bird with silver-gray flight feathers. The head appeared unfeathered — like the head of a Turkey Vulture. The feet seemed chalky white, excrement caked — also in the manner of the weak-footed Turkey Vulture. The flight was tipsy and wheeling, the efforts of a lightly ballasted bird struggling with the wind (which pretty much describes the flight of a vulture).

But if the dove had looked more closely, she would have seen the incongruities dismissed at first glance. The head of the slightly smaller, slightly slimmer bird in the group was not naked. It was broad, buteo-like. Only the large bill, yellow nares, and bare patch of skin before the eyes made it appear naked.

Neither were the feet weak and excrement caked. They were yellow with black-tipped talons. It was the broad white band bisecting the closed tail that projected the appearance of whitewashed feet and shanks.

And though the flight feathers did shine with the tarnished silver gleam characteristic of Turkey Vultures, closer scrutiny would have shown that the primaries and secondaries were finely barred — soot on white. At a distance, they only appeared uniformly silver-gray.

But the dove can probably be forgiven for having overlooked all of these things. In flight, even in direct comparison with Turkey Vultures, skilled field observers find it difficult to distinguish the birds, so closely do Turkey Vultures and Zone-tailed Hawks resemble each other, and the hunting hawk was not about to dispel the illusion. In search mode, the Zone-tailed's flight replicates the aimless, wheeling, nonthreatening flight of Turkey Vultures. Even after the hawk had sighted the dove and marked it for prey, it did nothing to distinguish itself as a hawk in vulture's clothing. It drifted with its confederates, waiting until the assembly had passed from sight before shedding its nonchalance and becoming eminently raptorial.

Had the dove been able to see the hawk once it broke from the group, she would have quickly seen through the disguise, seen . . .

How the bird turned back on a linear trajectory whose path intercepted the dove.

How the bird folded its wings in a long, shallow glide that would clear the top of the hill.

How the flight that had been unsteady was stiffened by purpose.

How the bird, when it had cleared the hill and made visual contact with its target, drew in its wings. Increased its speed. And in the final moments before impact, unleashed its feet, baring, finally, the predatory reality hidden behind the disguise.

But the dove saw none of these things. There wasn't time — not to note or to apprehend. There was only time to die.

Some have questioned whether Zone-tailed Hawks do or do not *actively* mimic Turkey Vultures. That they look like Turkey Vultures and fly like Turkey Vultures is uncontested, but whether the mimicry is intentional and whether Zone-taileds actually benefit from it has been the subject of debate.

Proponents of mimicry note that the range of the Zone-tailed Hawk — from southern California, Arizona, New Mexico, and Texas south to Central and South America — lies wholly within the more extensive range of the Turkey Vulture. Were Zone-taileds to derive no benefits from an association with vultures, it might be expected that Zone-taileds would occur where Turkey Vultures do not (but this is not the case).

Tacit evidence in support of mimicry is also found in the distribution and population density of Zone-tailed Hawks, because throughout their range the birds are not common. No more than a hundred pairs occur north of Mexico. Rarely do the birds nest within five miles of one another. If ruse is key to the bird's hunting success, it stands to reason that repeated disclosure would diminish its effectiveness. So like guild magicians jealous of their art, Zone-taileds keep the surprise in their act by keeping their distance and by limiting recruitment into their ranks.

Two eggs, deposited in a loose stick nest often positioned in a tall, solitary tree, are the norm. The typical nest tree is situated in a steep

The Wind Masters

canyon — against the cliff or close to the rugged mountain terrain that Zone-taileds prefer.

Here, where thermals and updrafts flourish, Zone-taileds navigate the crests of ridges, riding the currents on uplifted wings as vultures do. It was this affinity for slow, terrain-hugging flight that led to speculation that the Zone-tailed's supposed mimicry was, after all, no more than convergent evolution. If Zone-taileds flew like vultures, it was because the species' flight was aerodynamically suited for the habitat and hunting style.

But to accept this explanation is to ignore the fact that among buteos, only the plumage of juvenile Zone-tailed Hawks resembles that of the adults. In all other buteos, young birds have plumage that is distinctly (and often dramatically) different. Except for some white spotting on the breast and a barred (as opposed to banded) tail, adult and juvenile Zone-tailed Hawks look alike.

Morphological similarities between Turkey Vultures and Zone-tailed Hawks can be (and probably are) linked to aerodynamic considerations. Adaptation is often molded by multiple forces. But the patterning of a bird's plumage would seem to have little to do with its manner of flight. Clearly, likeness to Turkey Vultures holds some adaptive advantage if young seem to require the advantage, too.

Or looking at it another way, if a buteo, and a slow-moving buteo at that, is to take prey as agile as birds and as fleet-footed as a lizard, some tactical advantage is necessary. Why should it not be the obvious one? The one that makes a hunting predator almost invisible . . . until it is too late.

The dove whose life was about to expire had no opinion on the matter. Her understanding of the issue was no greater than her awareness. The last thing she knew, felt, and experienced was the impact of taloned feet that were extraordinarily large — wonderfully adapted for capturing birds in flight.

These, after all, were in the final analysis the evolutionary adaptations that killed her. All else is speculation. Although the dove could provide evidence, she could no longer offer testimony.

Elanoides forficatus
AMERICAN SWALLOW-TAILED KITE

IN AUGUST, ON THE PLAINS of South Carolina's Santee River delta, the day's heat starts early, building upon the heat left over from the day before. The sun starts red. Goes from morning gold to noonday white. It draws moisture aloft that coalesces into puffy clouds, which go through several stages of succession until by midafternoon they have matured into thunderheads that sweep all other things from the sky.

But the great splay-tailed bird of prey coasting along the forest rim paid little attention to the storm. White head gleaming, blue-black upper parts flashing with hidden light, she moved like scud over the bowed heads of trees, seemingly immune from wind or harm — and perhaps she was. After all, she was an American Swallow-tailed Kite. A bird whose mastery of her element is complete. If she could not outrace the storm, then she would parry its force away or turn it to her advantage, drawing energy from the tumult the way diplomats thrive on crisis. She said this with the precision of her movements. She said this with her easy, indomitable poise.

The bird, an adult female, was one of a small but stable population of fifty pairs localized in the vicinity of South Carolina's Francis Marion National Forest. This great coastal tract, stretching across interior forests and delta, offers secure haven for nesting kites and the canopy-topping trees that kites choose to hold their nests. The marshes provide the surfeit of dragonflies that are the bird's principal prey. And although the distances from the nests to the marsh are great (in some cases more than ten air miles), it mattered little to the bird. It was only distance, after all, mere open air. What is distance to the likes of a Swallow-tailed Kite?

She and her mate of several seasons had arrived in late March (although in Florida the birds may arrive weeks earlier). In the spring skies over the live oaks and loblolly pines, they had courted like wind given form, copulated like innocents, and built their moss-lined nest with sticks plucked from the trees of this coastal Eden. They produced two eggs, the average clutch size for Swallow-tailed Kites, and raised two chicks who survived the two great enemies of Swallow-tailed Kite nestlings: buffeting winds and marauding Great-horned Owls.

The chicks were fledged now. Except for their shorter tails, the young were virtually indistinguishable from the formally attired adults. Through trial and error they had learned the rudiments of being American Swallow-tailed Kites — learned how to seine the wind for prey; learned how to mount the superheated air until the wings that span four feet were drawn to mere slits in the summer sky; learned how to fly, for hours, without beating a wing or how with the crook of a wing to bend the wind to do a kite's bidding.

And then, like adolescents whose faith in themselves exceeds their ignorance concerning the challenges of the world, they had departed, taken the first steps on a long journey that would end on another continent. In a few days, the adult female and her mate would leave too, heading south to the west side of Lake Okeechobee, Florida.

There, in the dragonfly-rich skies near the lake, they would join Swallow-tailed Kites from across the bird's range, from the Carolinas, Georgia, Alabama, and, particularly, Florida — the state where the vast majority of North America's Swallow-tailed Kites nest. Once Swallow-tailed Kites ranged north along the watercourses of the Mississippi drainage and the riparian rim of the prairies. Their nests

AMERICAN SWALLOW-TAILED KITE

"She moved like scud over the bowed heads of trees, seemingly immune from wind or harm."

The Wind Masters

were found in twenty-one states. But logging and the agricultural conversion of natural lands compromised the birds' food and habitat needs; shooting reduced its numbers. By 1900, Swallow-tailed Kites were extirpated over much of their range. The remnant North American population, now estimated to number between 2,000 and 3,000 birds, was confined to seven southern states.

The post-breeding, premigratory assemblage near Lake Okeechobee begins to gather in July. At its peak, as many as 1,500 birds may gather — a number that may represent half the Swallow-tailed Kites to be found in the United States. Sometime before mid-September, presumably after their bodies have stored whatever fuel they need to meet the energetic demands of migration, the birds depart for South America. Some may follow the path laid down by the Florida Keys to Cuba to the Yucatán. Others may island-hop, following the curve of the Lesser Antilles, making landfall on the north coast of Venezuela. Still others, perhaps the majority, take the most direct path and simply vault the Gulf. The bird's ability to cross the Gulf of Mexico by powered flight is speculative, the evidence circumstantial. But the evidence supports such an assertion.

In the first place, it makes little sense from an energetic standpoint for a species to gather in southern Florida, a geographic dead end, and then take the longer, surer land route to South America that circumnavigates the northern rim of the Gulf of Mexico, as some authorities have suggested.

Second, hawk watching is an avid pastime in coastal Texas, a junction that kites migrating around the Gulf of Mexico must logically pass. Though large numbers of other raptor species have been seen in places such as Corpus Christi and Santa Ana National Wildlife Refuge — among them many thousands of Broad-winged Hawks, Swainson's Hawks, and Mississippi Kites — Swallow-tailed Kites are conspicuously absent in both spring and fall.

Third, in April, many observers have watched Swallow-tailed Kites approaching the Dry Tortugas and the coastal islands of Mississippi and Alabama from the south and continue on to the mainland. The birds show no sign of stress and no inclination to make an interim stop.

Finally, nobody who has ever watched an American Swallow-tailed Kite fly could possibly doubt that the birds are capable of per-

AMERICAN SWALLOW-TAILED KITE

forming any aerial feat. My respect for the bird's abilities is such that if it was postulated that Swallow-taileds spend the North American winter on the moon, instead of in South America, I would consider the possibility seriously before dismissing it — and it is not certain that I would dismiss it.

The bird continued to move along the tree line, weaving like a figure skater through invisible obstacles, making light of the storm's best buffeting efforts. Time and gravity seemed suspended. Friction could find no hold. The great blue-black bird with the sliver wings and deeply forked tail flew the way angels must in their dreams — but if an angel, then a killer angel; and if viewed from the multichambered eye of a dragonfly, then a nightmare and not a dream.

The bird stooped, fell — but not suddenly! Not abruptly. There was no halting check to the bird's momentum, no disharmony between the bird and the air that suspended her. The only thing that changed was direction. First the bird was moving horizontally, then she was not.

On half-folded wings she dipped toward the marsh, enveloping a large green dragonfly in Lilliputian feet. Then, arching her neck, the kite consumed the insect on the wing. It all happened so quickly and so effortlessly that the capture and repast had the quality of an afterthought. Only the wings of the insect, shivering earthward, showed that anything had happened at all.

To the north, across the marsh and still beyond the reach of the clouds, other raptors were aloft — a kiting Red-tailed Hawk, an Osprey laboring with prey, a Mississippi Kite with its wings drawn down to arrow-point angles, and a host of vultures walking windy tightropes in the sky. All were gifted fliers. All could boast a specialized finesse that distinguished them from other raptors and gave them firm standing among the ranks of the Wind Masters. None, in the limits of their skills, in the reach of their dreams, could approach the aerial finesse of the Swallow-tailed Kite.

Throughout human history, birds have won human admiration for their resplendent colors. They have been loved and celebrated for their song. But the attribute that birds have that humans covet most is their power of flight, their ability to put their wings to the wind and leave the earth behind.

The Wind Masters

It is not a difficult thing, flight — not in theory; not even in practice (so long as evolution has molded a creature to fly). Flight is a simple matter of playing two natural forces against each other: gravity and air pressure. The problem humans face when trying to understand flight is twofold. First, we tend to forget that flight does not occur in a vacuum. It occurs in a real, tangible fluid — air. Just because air cannot be seen does not mean that it is not real or that it is not subject to the laws of physics and nature.

Second, we cannot ourselves fly, at least not without extraneous, mechanical support (at least not for long). Sustained flight is quite beyond our physiological capacity, so we regard it with wonder and awe.

Birds are different. Birds are adapted for locomotion in air. Their skeletal framework, their structural support, is light and strong, the bones hollow, the articulated joints (in many cases) fused to anchor firmly the muscles that control wing movement and to protect internal organs from the crushing force of energetic flight.

The wings, which are specially modified forelimbs, serve as airfoils just like the wings of aircraft — convexly curved on the top, flat or slightly concave on the bottom. When knifing through air they create lift by forcing the air passing over the wing to cover more ground than air passing below. Spread thin, the air above the wing cannot push down with the same force as the air below pushes up. The net effect is lift, and the faster the air moves over the top of the wing, as opposed to below the wing, the greater that lift.

Feathers are perhaps a bird's greatest evolutionary adaptation. They are supremely light, flexible, strong. They shed water, trap heat, contour and conform to create a friction-shedding body suit. Certain specialized flight feathers, called primaries, are responsible for propulsion in powered flight — the type of flight used by most birds. Each primary is a miniature airfoil, a tiny wing unto itself. Passed vertically through air, it sets up the same airflow imbalance that occurs over the surface of a wing when air moves across it, *except* that with a downward-slicing primary the vacuum forms in front of the feather, not above. The result? The feather is pushed forward and the rest of the bird goes along for the ride.

There is another way that birds move forward, one that is particu-

AMERICAN SWALLOW-TAILED KITE

larly important to birds of prey. This is gliding. In this type of flight birds use gravity to pull themselves forward — but at a price. When gliding, a bird is always sinking relative to the air around it. Unless the air the bird occupies is also rising (as it does in a thermal, a rising column of warm air; or an updraft, a vertically directed flow of air), the bird will not be able to sustain its flight indefinitely. It will be forced to resort to powered flight, or it will be forced to seek out some upwardly mobile air, or it will have to land. There are no other alternatives.

Evolution has molded different species of birds to excel in different kinds of flight — each tailored as an advantage to that species in the habitat that it occupies. Some birds, such as woodland thrushes, are wonderfully adapted for direct, powered flight through forest. Some, such as gulls, are skilled soaring birds, able to seek out and take advantage of rising columns of air and ride these energy-saving elevators aloft. Other birds, such as swallows, are open-air acrobats, able to dart, twist, and accelerate quickly — qualities that help them capture insect prey in the open.

What distinguishes birds of prey from other birds is not necessarily the superiority of their aerial refinements. In fact and in truth, there are many birds on the planet whose powers of flight match or exceed the aerial skills of most raptors. A White-throated Swift or a pigeon will outpace a Peregrine Falcon in level flight. An albatross or a shearwater can find lift under conditions that would defeat the wings of a harrier or a condor. Most woodland songbirds can outmaneuver a Sharp-shinned Hawk in open country or wooded terrain, and hummingbirds can fly rings around them.

What distinguishes birds of prey from other birds is that, taken in sum, no group of birds has evolved that exhibits so many different modes of flight. Not only have birds of prey mastered the wind, they have done it in every conceivable manner — from the thermal-mantling flight of vultures, to the wind-cleaving dash of a falcon, to the fusion of wind and bird that is the flight of the Swallow-tailed Kite.

The storm was closer now. The great dark cloud loomed above the trees, enveloping the sky. Wind tore through the treetops and flattened the tall marsh grass, but in the lee of the forest edge the kite

The Wind Masters

found dragonflies in numbers and no reason to leave the sky. She moved along the tree line, rising and falling like a lancer's banner. Sometimes when prey was sighted her wings would gently convulse and the bird would know sudden speed. At other times she moved so slowly that it seemed she must fall from the sky, like a normal bird, but the Swallow-tailed Kite is not a normal bird.

Part of the kite's secret powers of flight are bound in its wings. Long and tapered, they enjoy the same high aspect ratio that puts dash in the falcon's flight and float in the sylvan form of the shearwater. But long, thin wings work best at high speeds. At low speed, eddies form along the trailing edge that interferes with the passage of air over the airfoil, causing airborne objects to lose the lift that holds them aloft. The result is a stall.

To compensate for their wings' structural shortfall, Swallow-tailed Kites bring their tails into play. Splayed widely, the split tail can be used as an extended wing — adding width to the airfoil and stability at low speed. By changing the shape of its tail the kite can transform itself from a falcon to a broad-winged soaring hawk, and more. Because each wing and each side of its swallowlike tail can be operated independently, a kite possesses the ability to micromanage the air it moves in; to master the wind in masterly fashion. Few other birds can do this. None, with the possible exception of the frigatebirds, can do it so well.

The first large drops of rain began to pock the marsh and the dragonflies quit the sky. The wind that had been shrill took on a note of menace and the horizon turned indistinct and gray.

The kite increased her speed, staying ahead of the rain for a time. Then the curtain seemed to catch her, and just before rain enveloped the marsh, it seemed that finally she perched.

But it may have been that the rain fell only between her and the rest of the world. It may be that the American Swallow-tailed Kite is as impervious to storms and rain as it is to gravity and friction.

And if this is so, then it might be that the storm passed her by or, more nearly, that she passed through. Slipping between raindrops as neatly as a kite in the breeze.

AMERICAN SWALLOW-TAILED KITE

TWENTY-EIGHT

Ictinia mississippiensis
MISSISSIPPI KITE

TWO MEN, DECKED OUT in brightly colored but otherwise casual attire, stepped away from their motorized cart and onto the green. The Mississippi Kite, sitting on her favorite perch along the windrow, had seen it many times before.

Teeing off, directing his attention to a small white object at his feet, one of the men raised a silver stick above his head and, bringing it down in an arching curve, sent the ball speeding through the air.

The kite had seen this, too.

At first the sphere ran true, straight down the avenue of grass. Then it started hooking . . . badly. Hitting the edge of the grass, the ball bounced gaily onward, disappearing into the wall of trees. Then the man with the stick became very vocal and very animate. Then he went for another ball.

The kite had seen this display as well, which is curious because on most golf courses, a golfer driving a ball into the woods will follow it up to see whether the ball is playable before accepting a penalty stroke. But not on *this* course, and particularly not on *this* hole. The windbreak flanking the fairway was home to a nesting colony of Mississippi Kites — among all the birds of the air one of the most graceful, and among all the hazards on the Oklahoma country club course, the most hazardous. Golfers who breached the territorial integrity of the colony had lost favorite caps (and, on occasion, even a little scalp!) and had learned to give the woods the widest possible berth.

The kite, one of several subadults in the colony, did not have anything in particular against golfers — no more than she did against Great Horned Owls, Cooper's Hawks, crows, Blue Jays, or other creatures too stupid to recognize the boundaries of a Mississippi Kite colony. Gentle in appearance and countenance, she and the other members of the colony were indifferent to people — so long as they left their nests in peace.

Interestingly enough, and a summer's surfeit of touchy encounters notwithstanding, had the golfer chosen to retrieve his ball, he could have done so with impunity. It was late August, almost September. The last nestling from the colony's eight active nests had fledged two weeks earlier. The kites had called a halt to hostilities for the season.

In the minds of Mississippi Kites it was time for flight, not fight. Migration was in the air; the great journey to the south was about to begin. It was something the female could feel in her genes. It was evident from the nervous agitation of the birds around her that they felt it, too.

Migration! It was her absolute favorite thing, she'd decided. Hours on the wing. New horizons. New companions. New heights — as regular as the next thermal, followed by long, long effortless glides.

She couldn't *wait* to surf the cloud streets over Panama again.

Last year, when she'd made her first journey to the grasslands of Argentina, it had been a *total* adventure — a young kite engaged in the great social exodus of her kind. The return trip, thermal-hopping along the meteorological battle lines drawn between winter and spring, had been a blast. Though she did not fully appreciate the urgency that drove the adult birds north, she was nevertheless moved by it.

It was when she reached the Oklahoma colony on the first of May, racing and beating a dark blue norther to the line of trees, that she had decided that migration had to be the greatest part of being a Mississippi Kite.

At least that's how she'd seen it then. Before the nesting season and the first full summer of her life. So *much* had happened during the ensuing four months. So many new and exciting things had involved her. She really couldn't decide (she decided) whether migration really was the best part of being a Mississippi Kite.

Or whether it was the challenges and social charges of summer.

Or the carefree days of austral summer, when all a Mississippi Kite has to do is soar and glide and seine prey from the sky.

She'd have to decide on her way south (she decided). If they ever started south. Their departure, to her mind, was overdue.

"Last year," she recalled, "we were all gone before now — mid-August. Last year we were under way long before this."

This wasn't entirely correct. The previous year, the colony had more or less disbanded and more or less relocated after the chicks had fledged — joining scores of birds from other colonies in Oklahoma, Kansas, and Texas in a premigration flock whose roost was some sixty miles away.

This year cicadas had gone through one of their cyclic eruptions and grasshoppers were likewise flush. With food locally abundant, there was little need and less inclination among the birds to relocate. In fact, Mississippi Kites from neighboring areas had been drawn to the town, using the trees around the memorial park for their evening roost. With food so abundant the kites built up fat reserves quickly — as fast as the birds could catch and consume prey. Even the young ones, on the wing a few short weeks, had little difficulty catching the large, slow-moving insects.

It was easy to distinguish the juveniles from the others perched in the trees. Warm brown backed where the adults were cold gray, streak headed and streak breasted where the underparts of adults were the color of silver in moonlight.

The subadult female was pleased to note how closely she resembled the older birds — had since spring. But to her continued distress, the second-year bird still bore vestiges of her juvenile plumage —

The Wind Masters

some unmolted brown feathers along her underwing coverts; the banded tail feathers of an immature bird that branded her a subadult. Not until she returned the next spring, sometime in April, would she brandish the flared black tail of an adult and the privileges that go with it — chief among them, the privilege of passing her genetic dowry on.

She'd already had a taste of the obligations that went with an all-black tail — even before she'd attained one.

Mississippi Kites may be territorial with regard to foreign intruders, but among their own kind, this southern raptor is eminently social. Mississippi Kites roost together, commute together, hunt together, and nest together. So casual are paired birds about their personal space that nonbreeding subadult birds are even allowed to attend young!

During the summer, the subadult had served as a "nest helper," assisting one of the eight adult pairs with their young. Over the course of 60 days between egg laying and fledging, she had assisted in the incubation, shading, brooding, and feeding of the two nestlings. She had also, and perhaps most importantly, served as a key component of the colony's nest defense system — an extra measure of fury to be thrown at trespassers.

This is why the young female knew so much about golfers — and why they, in turn, knew so much about her.

One of the reasons Mississippi Kites can afford to be so socially expansive relates to diet. Mississippi Kites are aerial predators whose primary focus is large insect prey such as cicadas, grasshoppers, and dragonflies. The normal abundance of these creatures, coupled with the irruptive and shifting nature of prey concentrations, coupled with the effortless aptitude kites apply to the capture of their prey, supports a relaxed attitude. Where other birds of prey defend hunting territories, kites can afford to share the air.

The focus upon aerial insect prey affects the lifestyle as well as the demeanor of Mississippi Kites. As a bird that is dependent upon temperature-sensitive prey, Mississippi Kites are among North America's latest-nesting birds of prey — as late as some Arctic species (even though the range of this distinctly southern raptor does not extend north of North Carolina, Illinois, or Kansas and west only to southeastern Arizona). In Florida, Georgia, Alabama, Mississippi, and Louisiana, Red-tailed Hawks may have young on the wing before Mississippi Kites have eggs in their somewhat unsubstantial stick nests.

Also related to diet, kites are late risers. The birds do not commonly begin to forage until the thermals that bear kites (and their insect prey) aloft are perking. Mississippi Kites can and do perch-hunt, making sorties from branches before the sun has warmed the air or when rainy weather puts a lid on thermal lift. The birds are also able to take vertebrate prey — including birds and reptiles — if need or opportunity dictates.

But airborne insects are the bird's forte, constituting virtually 100 percent of the kite's diet. Since open air is tied to no particular habitat type, Mississippi Kites are somewhat uncaring about the habitat dominating the earth below. Open skies over the southern Great Plains . . . open skies over southern forests . . . open skies over river bottoms . . . open skies over cities and towns . . . it's all the same to the Mississippi Kite.

All the birds seem to require to nest is a copse of trees surrounded by open, insect-rich space. In the past, on the Great Plains (the species' historical stronghold), this probably meant that the birds were closely tied to river systems and their associate riparian woodlands. Now, with the proliferation of planted trees — some in the form of town parks, some in the form of planted windbreaks — the birds are less restricted. In the last half of the twentieth century, Mississippi Kites have flourished and spread. Their populations, estimated to be between 25,000 and 30,000 birds, are stable or increasing.

More restless than hungry, the female left her perch and flew along the tree line, propelled by wing beats that were stiff and reserved. Folding suddenly, dropping abruptly, she plummeted to earth and, checking her fall, reached out with a foot.

She was already climbing when her talons folded over a seasonally fat grasshopper, one that had defeated the best efforts of groundskeepers.

Spiraling tightly, flapping occasionally, the bird brought her leg forward and her head down, tearing the morsel into kite-size pieces

The Wind Masters

"They climbed into an Oklahoma sky that is mile for
unencumbered mile the same sky as the sky over Argentina."

MISSISSIPPI KITE

that she bolted down. Dropping the remains, she flew out toward the semimanicured pond where dragonflies had been thick. She was not halfway to the pond before sighting prey.

Three quick, stiff wing beats added to her speed and an angled glide maintained it. Coming in behind and slightly below the unsuspecting insect, the kite abruptly leaped in mid-glide, absorbing as much as snatching the dragonfly with her foot. All that reached the ground were the insect's wings.

Her success, or perhaps their own restlessness, drew other kites, who joined her over the pond. As the sun climbed, as the birds fed, they were joined by more . . . then more. At first they hunted, climbing as their prey climbed. Then it seemed that they only climbed — moving with grace, moving without effort, moving in synergistic concert to some song that maybe only kites can hear.

They climbed until adults, subadults, and young were indistinguishable; climbed until the ties that bound them to the trees below were broken; climbed into an Oklahoma sky that is mile for unencumbered mile the same sky as the sky over Argentina.

Climbed until they disappeared.

It was a week later, at the bar in the clubhouse, that someone mentioned the absence of "the hawks." But it was two more weeks before anyone followed up a badly hooked ball.

208

The Wind Masters

TWENTY-NINE

Buteo platypterus
BROAD-WINGED HAWK

SEPTEMBER SUNLIGHT ROUTED THE MIST hanging over the eastern Pennsylvania landscape, falling upon fields, forests, homes, and rocky hillsides. Much of the light striking bare, open ground was absorbed; some striking vegetated areas was reflected back. Vertical inclines flush with the shallow rays of the sun received a full broadside of radiant energy. Shaded places and valleys still veiled in mist received little or no sunlight at all.

The result of this uneven apportionment was that some parts of the landscape warmed more quickly than others, and this caloric bias was passed, in turn, to the overlying air, which began to rise . . . and continued to rise because the air it encountered as it climbed was always colder.

These rising columns of air, called thermals, are the invisible elevators that soaring birds ride aloft. By using the sun's energy instead of their own, hawks can climb to tremendous heights. Then setting their wings, trading banked altitude for linear distance, birds of prey can travel many miles, expending little of their own energy in the

209

process. Some species, specially suited to capitalize on thermal lift, may even vault continents.

The crow-sized buteo with the broad white bands on his tail flew to the woodland edge and took a perch overlooking the power-line cut.

"Wasted energy," the hawk thought darkly. The bird, an adult Broad-winged Hawk, hated to resort to powered flight when there was so much free lift in the world.

It wasn't that the hawk was lazy, merely practical. A few weeks earlier, when the bird had been deep in the throes of parenting, its energetic output had been acute (and, as the four nestlings that had survived to fledging attested, successful). A normal clutch size for Broad-winged Hawks is two to three eggs. More than this, and the loss of one or more young to starvation is almost assured.

But the adult and his mate had risen to the challenge, aided by a nesting season blessed by a surfeit of prey — good numbers of frogs, good numbers of nestling songbirds, and (most important) a bumper crop of eastern chipmunks. Small mammals, ranging in size from shrews to rabbits, constitute the Broad-winged's principal prey during the nesting season. To the mind of the energy-conscious adult, shrews were too little gain for the effort; rabbits too heavy to heft. But chipmunks! Chipmunks were *just* right!

The nesting season was over. The young had fledged in August; their fate was no longer his concern. The nest site in a deciduous woodland in central Massachusetts was more than two hundred miles away — a full day's journey, as Broad-winged Hawks reckon distances. And prey, for the first time in a long time, was not the first thing on the bird's mind.

It was September 14. From Alberta to Nova Scotia, one of North America's most abundant diurnal raptors, the Broad-winged Hawks, were going south. Over much of the birds' route, the movement is a sweeping flood. But along the northern shores of the Great Lakes, and the New England coast from Massachusetts to Connecticut, the flood is constricted to the dimensions of a river, a river of raptors.

As the adult Broad-winged Hawk waited, he noticed other Broad-wingeds perched along the woodland edge — members of the troop he had journeyed with the day before; birds who had "fallen out," just as he had fallen out, when the day's flight had ended. Some of the hawks sat high along the slope of the ridge, some low. Most were ruddy-breasted adults like himself, but here and there he could see streak-breasted birds-of-the-year. As in most migratory raptors, the migration of young birds precedes that of adults. Juvenile Broad-wingeds (along with unattached subadults) had been moving since early August.

For Broad-wingeds heralding from the farthest reaches of the species' breeding range, the Maritime Provinces of Canada, and traveling to the southern limits of their winter range, the northern portions of Brazil and Peru, the journey would exceed 5,000 land miles. Though the air-mile distance between Nova Scotia and Peru is considerably less than this, it is of little avail to Broad-wingeds. To vault continents Broad-winged Hawks depend upon thermal lift. Since open water is not conducive to thermal production, the birds are forced to bypass the Gulf of Mexico and navigate the longer, Broad-winged–friendly land bridge of Mexico and Central America.

The Broad-winged's reliance upon thermals also explains why the adult Broad-winged Hawk was sitting, waiting for the sun to do its work. It made energetic sense. But patience is not universally trait-apportioned among migrating buteos — not even buteos whose bloodlines harken back to generations of long-distance champions.

Below the adult, a young Broad-winged left his perch and began flying toward the base of the ridge — a forested monocline that was the region's dominant geographic feature. Dubbed by Native Americans "the Kittatinny" (the Endless Mountain), the ridge extends from New York State through Pennsylvania. For many less thermally adept birds of prey, the ridge serves as a "migratory leading line." Updrafts cresting at its rim offer birds energetically friendly lift, and the geographic formation serves as a pathway in the sky.

At the base of a talus slope, where morning sunlight glistened off the face of weathered shale, the impatient youngster began turning circles — a maneuver he could manage only by augmenting his efforts with bouts of flapping.

"Amateur," the adult concluded. "Kid will run out of gas before he crosses the Rio Grande if he keeps wasting energy like that."

A veteran of three round trips to the tropics, the adult knew all about the rigors ahead and how to stack the odds in his favor. There were three simple rules he followed.

"If there isn't any lift, sit."

"If there's an opportunity to eat, eat."

The kid was breaking rule number one and muffing any chance to capitalize on rule number two. The adult, being more disciplined, had also been more fortunate. He had killed and eaten a chipmunk shortly after dawn, though he might have gone without — would often go without in the days ahead. For migrating Broad-wingeds, meals are, literally, catch as catch can; less important than putting miles behind. Other species are different, feeding regularly and sometimes daily during migration. But fasting, particularly in high-traffic areas where half a million birds may pass in a short span of days, is the norm for Broad-wingeds.

There was one final rule that was part of the adult's migratory formula for success, a complex one that was simple only in its expression: "Fly smart." By this the bird meant stay with the group, watch the others, capitalize on whatever opportunities you can.

Flying smart was *not* the same as "be a team player," which Broad-winged Hawks are not. Although the birds migrate in tremendous numbers — sometimes groups of 10,000 birds or more — these assemblies are aggregations forged as much by circumstance as by design. They are not, in other words, a "flock" except in the loosest sense of the word.

At the onset of a day's migration, at its close, and even during migration itself, Broad-winged Hawks behave as independent organisms. The celebrated concentrations of birds seen at migratory junctions such as Hawk Ridge in Duluth, Minnesota; Hawk Cliff, Ontario; Montclair, New Jersey; and Light House Point, Connecticut, are the result of masses of birds being confronted and diverted by water barriers. Concentration points along the Texas and Louisiana coasts and (particularly) at the constricted junctions of Veracruz, Mexico, and Panama work the same way.

Between the Great Lakes and Middle Atlantic states and the Gulf Coast, there are no formidable water barriers. The mass of migrating birds is free to spread out into more diffuse and numerically reduced

The Wind Masters

groups of birds, which go largely unnoticed until they strike the Gulf Coast and condense into a river of raptors once more.

Though not flocking birds in the social manner of geese or shorebirds, it is nevertheless apparent that Broad-winged Hawks do form groups, both large and small, and that these groups are advantageous to the migrating birds. Why else would so many members of an abundant species elect to move all at once along a limited number of routes? And if this strategy did not work, why, after using such a strategy for many years, are Broad-winged Hawks still among the most numerous raptors in North America?

The energy-conscious adult continued to watch the young Broad-winged Hawk's efforts, noting that although the bird was not rising, he was not sinking either.

"Kid's got something," he assessed. "Just doesn't know what to do with it."

Another adult must have reached this same conclusion — or perhaps the hawk had simply sensed the subtle change in the air. Gliding down the ridge, using energy he'd banked the day before, the bird came in just below the struggling youngster . . . fanned his wings and tail . . . circled once . . . flapped quickly and doubled back . . . then turning, opened his wings and tail to their fullest extent. Almost immediately, the adult began to climb.

"Missed it the first time," the perched adult noted with a professional eye. "Got it now, though." Even the young bird, having clued in on the example set by the other adult, seemed to be doing better.

Leaving his perch, gliding down the ridge as the other adult had done, the energy-conscious Broad-winged flew to a point just below the circling pair. Feeling air pushing up on his left wing, he turned that way, finding the lift he sought. It wasn't a strong thermal, and he didn't expect it to be. But it was strong enough. Turning tightly, staying within the rising column, the adult overtook the young bird (who kept overflying the edge). Noting that the other adult had turned north, toward the ridge, the three-time veteran followed when the thermal's energy played out.

Tails narrowed, wings drawn down to the shape of paring knives, the two adults reached the crest of the ridge in a tie. Though the other

BROAD-WINGED HAWK

adult had left the thermal first, the energy-conscious adult had been able to squeeze more altitude out of it. Trading this for a harsher glide angle, he had achieved greater speed . . . and a tie.

Simultaneously the birds flared wings and tails, presenting the fullest possible surface area — the avian equivalent of "spreading sail." There was no thermal on the ridge, but there was, as the birds suspected, a slight updraft spawned by a north wind — all the punch that was left in the cold front that had passed through New England two days earlier, triggering the exodus of birds. The two adults used the deflected wind current the same way they would a thermal — circling in the upwardly mobile air; gaining easy lift.

They might have chosen to ride the updraft along the course, at least until thermal production improved. They could have used the deflected winds to stay aloft and the tug of gravity to move them along, the way other species, including several Sharp-shinneds and an Osprey, were doing.

If the day had been cloudy, thermal-poor, they might very well have used the ridge-top escalator instead of looking to a thermal elevator. But it wasn't cloudy. In fact, with moderating winds and clear skies it promised to be a superb day for soaring.

Besides, though the ridge ran generally northeast to southwest (the direction they were heading), it wandered. The stretch they had intercepted was due east-west — a geographic digression not to the liking of the energy-conscious Broad-winged. When the pair spotted a dozen Broad-wingeds wheeling over the base of a fire-blackened bluff a mile south of the ridge, they set their wings and struck off cross-country to join them, leaving the ridge and its updraft to raptors less thermally adept.

All along the ridge and out over the valley, other Broad-winged Hawks were spiraling aloft — pairs, trios, and small groups mantling the first weak thermals of the morning, setting off in search of stronger ones. It was easy to pick out the thermals with the right stuff. Strong thermals boosted birds quickly aloft. Quick-climbing birds attracted other birds, who, by strength of numbers, attracted more birds still. By the time the pair reached the bluff, the original group had developed into a flock of nearly forty birds — or, as such groups are referred to in hawk-watching circles, a "kettle."

The Wind Masters

DAS

"*Simultaneously the birds flared wings and tails, presenting the fullest possible surface area — the avian equivalent of 'spreading sail.'*"

BROAD-WINGED HAWK

The term is descriptively apt. It refers to the visual impression projected by birds climbing in a thermal — a convolution that recalls water boiling in a kettle. Though the movement of swirling birds seems chaotic, it is not. There is rhyme and reason governing the movement of birds in a kettle. There may even be organization.

Within thermals, lift is not equally apportioned. The greatest vertical rise occurs near the core, the weakest at the edge. By observing the other birds in the kettle, individuals can position themselves to best advantage — saving time in their rate of climb; maximizing energetic profits.

Like finding thermals, being attentive to the climbing fortune of others and then riding the coattails of their advantage is another reason why migrating in flocks is more energy efficient than flying solo. It is, to use the tenet espoused by the three-time veteran, "flying smart."

The pair of adults had ceased to be a pair by the time they entered the kettle, passed through the growing ranks of birds, and joined the line of birds streaming southwest. They had become part of the group, members of an ad hoc flock. A quarter of a mile ahead, the vanguard of their group was spread across the sky, several bird lengths apart, searching for the next thermal.

Soaring birds locate thermals in several ways. They can find them visually by noting the telltale ascent of other soaring birds or by seeing airborne debris trapped in a thermal. Hawks can also locate thermals by tacking toward the windward side of forming cumulus clouds — the visual and terminal stage of a thermal.

Yet another way hawks find thermals is by being sensitive to the buffeting brush of rising air. Unfortunately, there is a lot of sky enfolding the world, and thermals are not evenly distributed. Two spaced birds, flying a parallel course, are more than twice as likely to randomly encounter a thermal as one. Three birds fanned across the sky are more likely still.

There were, in fact, six Broad-wingeds spread in an interception line at the head of the streaming group. The lead birds traveled half a mile before two birds on the right side of the line abruptly wheeled and began circling. Those birds strung out behind adjusted their course to fall in below them.

The Wind Masters

It was a good thermal. A real pressure cooker of a column. Birds entering the elevator rose quickly, towering above newly arrived birds. Another group of Broad-wingeds, more than a hundred strong, streamed in from the north to share the spoils. By the time the energy-conscious adult reached the top of the funnel, closed tail, trimmed sail, and started his glide, he found himself as much the leader of one migrating group as the follower of another. All along the migration route, from horizon to horizon, in a line running northeast to southwest, there were churning cyclones bound by umbilical cords of set-winged birds.

"Almost as good as the cloud streets in Panama," the Broad-winged thought, recalling the passage through the corridor of tropical cumulus that mark the contour of the peninsula. In fact, and very soon, the migrating Broad-wingeds were breaking through the base of newly formed clouds. The altitude was 5,000 feet. The time was 11:30.

As the day wore on, as thermal production peaked, the cyclones grew farther apart and the streaming glides lengthened. With lift so easy to find and so widespread, the mass of migrating birds became more diffuse. More a midday movement of traffic along some interstate than a rush-hour crush.

It was only after midafternoon, when thermal production began to taper off, that the pragmatic advantages inherent in flocking brought birds together once more. Though still impressive, the number of kettling birds was not as spectacular as the armada that had assembled during the morning. The masses of Broad-wingeds shunted east by Long Island Sound had found room to disperse over the great flat farmland of southern Pennsylvania. Distribution had also been affected by skill. Those birds who knew how to fly right had been able to cover more distance than those who were still fine-tuning their migration skills. As the sun's rays grew shallow and thermal production diminished, more and more birds were forced to "fall out" and seek shelter for the night.

Shortly after five o'clock, somewhere just over the West Virginia border, a group of thirty Broad-wingeds, all adults, felt the last thermal of the day slip away from them. They settled to earth, some to one side of a Piedmont valley, some to another. The adult Broad-winged who shepherded his energy so carefully glided toward a perch on the east-facing hillside. Though done as skillfully as any Broad-winged might, it still cost the bird two series of flaps to gain the perch he wanted.

He'd traveled nearly 240 miles. It was the first time he'd used his wings to do more than glide since leaving the power-line cut that morning. He was well on his way to completing his fourth trip to the tropics.

DAS

THIRTY

Accipiter cooperii
COOPER'S HAWK

SHE-AND-HER-BROTHERS WOKE with a start, but unlike the several other times the young Cooper's Hawk had been brought to wakefulness during the night, this time she remained alert. The first thing she noted was that there was no imminent danger. The second, that there was no prey. The third, that the rushing sound that surrounded her roost site had grown in volume (like the stream near the nest after the rains).

But it was not the sound that had wakened her. It was the light growing in the eastern sky and, of course, her hunger.

As the early bird catches the worm, the raptors that prey upon early birds must rise even earlier. Even in migration, the phase of her life that She-and-her-brothers was engaged in now, Cooper's Hawks are active early. Up before the sun. Foraging at first light. Seeking the prey that will fuel the energetic needs of the day's flight or, if you are young, unskilled, and unsuccessful as She-and-her-brothers had been the previous evening, meet the energetic deficit left over from the previous day's flight.

219

COOPER'S HAWK

She-and-her-brothers was a fairly large bird, nearly 460 millimeters in length (longer than an average Red-tailed Hawk) and 500 no-fat grams (almost as large as an average, healthy Red-shouldered Hawk). To remain healthy, the bird needed to ingest about 70 grams of food per day — the numeric equivalent of seven Black-capped Chickadees, six Black-and-white Warblers, two or three Gray Catbirds — or one robin. Insofar as it requires less time and energy to capture one robin-sized bird than six warbler-sized birds, Cooper's Hawks favor larger prey, birds up to the size of pheasants, and mammals as large as hares.

But She-and-her-brothers was too hungry to be choosy. A catbird or a chipmunk would have been just fine. With a departure so quick and so silent that even the branch on which she perched remained ignorant, She-and-her-brothers departed. One moment she was there, the next she was not. She had become a brown-cloaked, streak-breasted torpedo, weaving through a world of shadows.

Hunger! It was the pole that moved the needle of her compass, the concern that made all others inconsequential, and it seemed her short life had been little more than an applied study in starvation. It had started in the nest, a low-profile platform of sticks resting atop the junction of five branches that radiated outward from the trunk of a fifty-foot white pine. There she and her four siblings had begun the journey that would be their lives. Only two would leave that nest; the other three would carry on by proxy.

It was not the fault of the parents that they could not secure enough prey to meet the needs of their young. Both of the blue-backed, orange-breasted adults were capable hunters and attentive parents (though only the female had raised young before). Their April courtship in the skies over southern Quebec had been spirited, complete with morning dueting, aerial displays, and great fuss made over the location and building of the nest.

The five eggs, white with a suggestion of blue, were laid on succeeding days, but incubation was delayed until the third egg was laid — a strategy to reduce variation in the size and age of nestlings. Early-hatching young are more developed than later-hatching nest mates. In the competition for food, these older siblings dominate. By retarding the development of first-laid eggs, adults can compress the

The Wind Masters

*"She had become a brown-cloaked, streak-breasted torpedo,
weaving through a world of shadows."*

221

COOPER'S HAWK

hatching period from five days to three, thus blunting the competitive edge afforded older chicks and bettering the chance of younger birds surviving.

As fortune had it, two of the first three eggs laid produced females, which even as nestlings are larger and stronger than males. To make matters worse, food was not plentiful. The male, who did virtually all of the hunting during incubation and the first three weeks of chick development, was hard put to meet the food needs of his mate, their young, and himself.

Sometimes traveling a mile or more from the nest, the male combed the countryside searching for medium-size songbirds such as robins, Blue Jays, and thrushes; small mammals, including rabbits, gray and red squirrels, and especially (maybe particularly) chipmunks. Rabbits and Ruffed Grouse were at a low in their population cycles. Neotropical migrant populations were faltering; year-round residents were moderately depressed by a brutal winter.

Whereas in good times the male might carry prey to the nest seven or eight times in the course of a day, this nesting season his efforts seemed able to muster no more than three or four successful forays per day — half the amount of food that the nestlings required. Never going directly to the nest, he would fly to one of several nearby branches and call, and the female would leave the nest to retrieve the prey.

Tearing off small pieces, she offered them to the bloom of open-beaked young; the red of freshly torn flesh stimulates chicks to feed during the early stages of nestling development. Chicks go for each offered morsel the way the bull goes for the matador's cape. Inevitably it is the largest, the strongest chicks who command the lion's share of food.

Out-muscled by his siblings, growing steadily weaker, the last male hatched died in the first week of his development. The next younger sibling, still wrapped in fluffy white down, died two days later. Both were immediately fed to the other nestlings by the mother.

To most people, particularly those whose moral and ethical standards have been conditioned by any one of the earth's dominant religious cultures, the act of cannibalism is appalling; the idea of a mother butchering and feeding an offspring to her other young is monstrous.

The Wind Masters

But our regard for cannibalism is shaped and maintained by the ease of our conditions. In instances when humans have suffered the severe nutritional privations young birds of prey suffer as a matter of course, even our high moral standards have surrendered to desperate need.

From a survival standpoint, cannibalism makes perfect, pragmatic sense. Nature is profligate as well as practical. She almost always produces more offspring than are necessary to maintain a species — whether it is herring in the sea or Cooper's Hawks in the nest. Perhaps 80 percent of all Cooper's Hawks hatched in any given year fail to see a second. From a species standpoint, and from the standpoint of a parent striving to pass on its genetic dowry, if the flesh of a sibling can stave off total nest failure, it is an option worth taking.

Certainly the world's Christians, who practice a faith founded on the selfless sacrifice of the Son of the Creator, whose flesh and blood must be consumed in order for practitioners to attain salvation, must agree.

The female Cooper's Hawk was not a Christian, but neither did she suffer the emotional trauma that a human mother might face in similar circumstances. At the point her young died, her attachment did likewise. They had become food.

She was not at the nest when the last of her male offspring died. The incident occurred early in the chick's fourth week of development, after the female had joined the male in hunting. The chicks, now almost entirely feathered, were old enough not to need constant care, and the weather had turned rainy and cold, making hunting even less productive than before.

Nevertheless, She-and-her-brothers, her sister, and the surviving male were wet, cold, and famished. The rain and short rations had brought the nest close to failure. All the parent birds had managed to bring in during the last three days were a Hermit Thrush, a Blue Jay, and a White-throated Sparrow — the combined energy value of which would hardly have satisfied the food needs of a single young Cooper's Hawk, much less three.

Near midmorning, the male arrived with a chipmunk. Finding the female absent, he went directly to the nest, dropped the rodent, and fled as the nestlings grappled for the offering.

The birds were large enough to move about freely and large enough to inflict damage with their feet. Both females fought for possession, each trying to wrest the animal from the other. The male, weakened almost past the point of caring, was not so much involved in the struggle as enveloped by it.

Somehow, She-and-her-brothers was upended in the struggle and toppled to her side. Somehow, in her flailing efforts to right herself, her foot found the head and neck of her weakened nest mate.

The young male struggled, as any animal might. She tightened her grip, as any Cooper's Hawk feeling life in her prey would. It wasn't a long struggle. In fact, it was short. Her grip was strong, the talons mortally fixed, and the life in her sibling eager for release.

Blood seeped from around the embedded talons and mixed with the rain, staining the male's neck red. Seeing it, the other female reacted instinctively, reaching in with a bill, tearing off a piece of flesh, bolting it down, then reaching for another. By the time the very wet and very tired adult female returned to the nest, a red squirrel clutched in her talons, the fratricide victim, her offspring, was nearly a third consumed. Seeing that her young were satisfied, she ate the squirrel herself.

The rain lasted two more days. The body of the male did likewise. Once the rains abated, hunting improved, and one week later, when they were about 35 days old, the surviving sisters, more than once sustained by the flesh of their brothers, fledged. They remained in the company of the adults for two more weeks, growing more self-sufficient every day. Near the end of July, She-and-her-brothers made her second kill. That night she did not return to the nest site.

She wandered without direction. Foraging more or less continuously. Feeling the pinch of hunger almost constantly. By September her wandering had taken a distinct southern tack. By early October she was heading south, through the mountains and valleys of New Hampshire . . . across Massachusetts, Connecticut, the corner of New York, and then into New Jersey. It was here that she had spent the night.

Surprising nothing, She-and-her-brothers took a perch at the edge of the long woods, overlooking what appeared to be a black river, flanked on either side by short-cut grass. The river was crowded by large, fast-moving creatures with glowing eyes, but the hawk's eyes fastened not on these but upon the many robins, migrants like herself, feeding in the grass beyond.

There was no cover beyond the woods. Her only hope seemed to be a quick, all-out dash across the rushing stream. She left the perch without a sound, accelerating rapidly, focusing her attention upon the birds in the flock.

She passed directly in front of a minivan, was picked up by the slipstream passing over a sport sedan, which turned her sideways, slowing her attack and saving her from certain collision with a pastry delivery truck. She passed behind the truck, still unseen, and was upon the robins before they could flee.

She was lucky to be alive and she was lucky to have prey, but She-and-her-brothers had the luck and the fortune of four. If her luck held, if she lived to maturity, she would win the prize of all living things — the privilege of passing her genetic inheritance on. Her brothers had given her a head start toward this end. The rest was up to her.

THIRTY-ONE

Buteo lagopus
ROUGH-LEGGED HAWK

THE DORTHEA, HALIFAX BOUND, was 140 miles south of Cape Sable, in calm seas, when navigator Chris Erikson left the bridge to have a smoke. The cold wrapped itself around him like a wet cat.

After fifteen years in the merchant marine, Chris was familiar with both the cold and the sea. Impervious to the former, in love with the latter, he had his childhood on a farm in northern Kansas to thank for both his fortitude and his profession. Like many people living in landlocked places, the farm lad had developed a romantic infatuation with going to sea.

The *Dorthea*'s navigator was halfway through his cigarette when he noticed the bird flying low over the water, approaching the bow. It was about the size of a large gull — but broader, bulkier, and it didn't fly right. The wing beats were labored, not languid, and they moved with a rowing motion.

"Like an oarsman," the thought came to him. And the bird, moving steadily across the flat, windless sea, did recall the tireless efforts of some long-limbed Viking oarsman.

For a moment it looked as though the bird and the ship were going to collide. Then he lost sight of the bird as it dipped below the mass of steel. By the time he relocated it, the bird was starboard and receding fast — a large, long-winged shadow of a bird with white patches bleeding through the wing tips and a bright white tail tipped with a black swath. It looked, he realized, like a hawk.

"What the devil would a hawk be doing out here?" he wondered, looking around for a landmass he knew did not exist. "*Must* have been a gull," he concluded, and by the time he finished his cigarette the encounter was forgotten.

If anything, the hawk gave even less thought to the encounter than the puzzled seaman (and it is debatable whether the bird consciously noted the ship at all). The hawk was very preoccupied at the moment. It was fighting for its life, and this is a very consuming ambition.

The bird had been flying steadily for nearly eight hours. It was close to exhaustion. It was holding itself just above the reach of the swells. And it was doing its utmost not to be mastered by . . .

Pain. It cramped in every muscle. It burned in every nerve. It strove to make the bird oblivious to everything but . . .

Pain. It was like an ocean above the ocean through which the bird swam, and it seemed to the bird that with every wing beat, it dipped its wings into torment and pulled itself through agony.

The bird, a young Rough-legged Hawk, had left the coast of Nova Scotia shortly after sunrise. Over open water, with no updrafts to support it or thermals to bear it aloft, the bird had no choice but to employ active flight to carry it — a mode of transport three to six times as energy consumptive as soaring flight.

Why had the bird elected to make a dangerous water crossing instead of taking the safer, easier course that led back along the length of Nova Scotia to the North American mainland? Because the bird had assessed the merits of both courses of action, factoring in such migration-related variables as time, distance, and weather conditions, and concluded that the crossing was worth the risk.

ROUGH-LEGGED HAWK

Its decision had also been influenced by its health, which was good; its energy reserves, which were adequate; and its experience with water crossings, which, to this point, had been positive.

There was one more factor that figured in the bird's decision, a salient one. The bird had no idea how far it would have to travel to find land. It had its youthful inexperience to blame for not according this variable more weight.

For the first five hours of flight, its youth and its strength had sustained it. It had held to an altitude of 200 feet above the sea. It had maintained a flight speed of about 20 miles per hour.

But each wing beat had a price and every mile had taken its toll. Now, after 140 miles of steady flight, it was tired and it was struggling. Now the bird had little to combat the pain and the fatigue except THE GLIDE and *the dream*.

The GLIDE. The merciful set-winged pause that followed each series of wing beats. It was akin to the rest that the heart takes between beats and it was real.

The *dream*? The dream was like the trance that marathon runners experience to buffer their pain. It was something that grew out of the rhythm of the bird's wings, grew as the pain grew. And although the dream was not as real as the glide, it was as real as the pain.

In the dream there were . . .

siblings; and feedings; and feathers [that were]
warming; and painless; and . . .

G   L   I   D   E

In the dream it was . . .

softness; and effortlessness; and safeness; [it was]
nest; and rim; and sky . . .

G   L   I   D   E

nest; and rim; and sky . . .

After 23,000 wing beats, it was the dream, as much as the glide, as much as the Rough-legged Hawk's powers of flight, that kept it moving through the ocean of pain and above the reach of the waves.

The Wind Masters

The nest that was part of the dream sat on a stony bluff near the mouth of the Alexis River, not far from the Labrador Sea. The rim that was also part of the dream was made of spruce and willow sticks, which is typical of Rough-legged nests, and the sky was the sky that looked down upon the Canadian province of Newfoundland.

Sometimes the sky was blue. Mostly it was gray and fogbound and the wetness of it surrounded the four nestlings like the pain that now enveloped the bird. Between cold and hunger was the parent bird — the one whose brooding warmth was like safety itself and whose lemming-rich bill took the hunger out of the world. These were the elements of the dream. And they were among the earliest memories in the bird's short life.

It is not at all uncommon for birds of prey to migrate over open water, although most species will risk only short crossings — less than 20 miles. But there are some species that habitually make water crossings in excess of 100 miles: species such as the Northern Harriers, which navigate the 110-mile crossing to Cuba, and the mass of American Swallow-tailed Kites, which vault the Gulf of Mexico en route to and from winter territories.

But even these migratory leaps of faith pale in comparison with the over-water migrations conducted by such powerful fliers as the Gyrfalcon and the Greenland Peregrine, who face at least a 500-mile crossing to reach the North American mainland.

There is even reason to believe that the adult male Peregrine may be pelagic, or nearly so, in its autumn migration; that birds may jump off from points in northeastern Canada and not light until they feel the strength of a tropical sun upon their backs. The minimum distance of such a flight would be about 2,000 miles. Fantasy? Perhaps. But the Amur Falcon of India, a bird that bears structural kinship to the Peregrine Falcon, makes a comparable crossing to the west coast of Africa twice a year.

So the situation that the young Rough-legged Hawk found itself in was not uncommon among raptors. It was, however, uncommon among buteos, a genus that relies very much upon the elevating advantage afforded by thermals and updrafts to aid the migratory passage. In fact, among all of North America's buteos, only the

ROUGH-LEGGED HAWK

Rough-legged Hawk seems moderately unfazed by long-distance, over-water crossings. Its wing loading, comparable to that of the Osprey, another over-water migrant, makes this possible. Its occupation of nest sites on islands scattered throughout the Arctic seas makes it unavoidable (at least in autumn, when the islands and the Canadian mainland are not knit by ice, as they are in spring).

Oblique testimony to the bird's ability to vault aquatic barriers can be seen in the scope of the bird's range. Rough-legged Hawks are Holarctic, distributed across the tundra regions of the Northern Hemisphere, a boast no other buteo can make. But the highest tribute to the bird's powers of flight may be its impressive record of vagrancy. Rough-legged Hawks have been recorded on the islands of Bermuda and Hawaii. And there is only one way for Rough-legged Hawks to reach such distant outposts. That is to . . .

reach. And reach. And reach. And reach. And . . .

G   L   I   D   E

Reach. And reach. And reach. And reach. And . . .

G   L   I   D   E

Up, and out, and down, and back. Up, and out, and down, and back. Row, and row, and row, and . . .

G   L   I   I   D   E

It was the rhythm of the bird's life, and the horizon was the focus of the bird's life. Bird was the subject. Horizon was the object. In the dream, and in survival, they were one.

It was the horizon that had actually unlocked the dream in the bird. Near midday, when the bird's strength was starting to falter and pain had begun eroding the bird's resolve, the horizon had offered the bird a focus beyond the pain, a target, an escape.

The horizon that was

gray, and green, and open . . .

G   L   I   D   E

Open, and inviting, and familiar . . .

The Wind Masters

Open, familiar, and comforting . . .

G L I D E

It was like the horizon the bird had known from the rim of the nest, and later from the vantage of its wings as it soared over the tundra and taiga reaches of Labrador. In the empty sea, in a growing ocean of pain, the horizon invited the reach of the young Rough-legged Hawk's wings just as it had the day the young hawk had challenged the world beyond the nest and won.

Minute-by-mile-by-wing-beat the horizon had become less the rim of a hostile sea and more the border of familiar haunts. More and more the bird flew less in the real world and more in the dream.

Even the dream was somewhat familiar, and this too was comforting. The bird had fallen into the rhythm of the dream when it crossed from Newfoundland to Nova Scotia — a distance of some 70 miles; a distance the bird had navigated without incident. An earlier crossing over the Strait of Belle Isle, a 20-mile span, had been too short to conjure the dream; had been no effort at all.

The voyage the bird was engaged in now was greater than the span of Cabot Strait. Far greater, not even finite, because the bird had no way of knowing how far it needed to travel. All it could count on was the genetic promise, evidenced by its existence, that other Rough-legged Hawks had made perilous leaps of faith before and survived. All it had was all that it had inherited. And the strength it had left.

And the dream.

As the bird's exhaustion increased, the range of its focus decreased and the dream diminished in measure. The horizon grew misty — like the Labrador coast when the fog rolls in from the sea. And the focus of the dream became the things the bird recalled before its wings ever reached for the horizon; the things it could see from the nest.

Hummocks, and hilltops, and high-soaring . . .
ravens G L I D E
tundra and taiga and stone

"All it had was all that it had inherited. And the strength it had left."

The Wind Masters

Two more hours passed and now the bird dreamed of the world before the horizon; the world that consisted of itself, and three nestlings, and a stick-rimmed nest. This was the stage of the dream that the bird was experiencing when it had encountered the *Dorthea,* and why the bird did not react to the ship except in its dream.

In the dream, the bulk of the ship had seemed like the parent that appeared at the rim of the nest. It came with lemmings. And it tore them into morsels that hung before eager open mouths. Just within reach.

Bright, and red, and dangling
G   L   I   D   E
red, and red, and good,
G   L   I   D   E
good, and good, and gone,
G   L   I   D   E
gone, and soft, and dark
G   L   I   D   E
soft, and dark, and dark

The dream after the sun went down was like the world beneath the brooding parent, the world before feathers and flight.

There was a weight pressing down on the bird. It felt it pushing down . . . down. . . . The bird wanted to come up. To fly above what it could see beneath the Pain. But it could not. It could not. There was so much Pain. Pain that was greater than the dream. And the only thing that seemed greater than the Pain was the Darkness beneath the Pain that promised

softness and safeness and
R   E   S   T
painless and effortless
R   E   S   T
dark . . .
andwarmandpainless
REST
dark and . . .
warmand . . .

silent
REST . . . *rest* . . . *ressst*

Its wing beats were arrhythmic. Its glides abbreviated, almost nonexistent. It was flying dangerously close to the water, pressed down by the growing weight on its wings. And the water, swelling to reach it, was dark. Like the sky. Like the world before the struggle against the barrier to the world had been broken and the bird had emerged, still wet from the egg; too exhausted to lift its head. It had taken all the energy it had to break through that barrier. But it had broken it. It had. Because it was a Rough-legged Hawk. And a survivor. It had come from a long line of survivors. It . . .

. . . Had begun with a crack! A point of light that had erupted before its eyes. That cut through the opaque darkness that surrounded it. That led to the world beyond.

A point of light. Like the one . . . NOW. That cut through the darkness that was cutting the bird off from the world . . . NOW. A thing that was something other than pain. A single . . .

Light . . . light . . . light . . . that came and went like a wing beat; that interrupted the bird's passage down the long, dark glide that ends in nothing. It might not have been real. There was only one way to know if it was real. But it was all there was except for darkness.

The bird's strength was gone. The horizon was gone. Even the dream was gone because the pain that sustained the dream was also gone. It was being absorbed by the darkness.

So the bird that had no strength, and was too exhausted to feel exhaustion, flew toward the light that was sometimes there and sometimes not there. It flew because its wings had forgotten how to stop, and because the bird had come from a long line of survivors who had made great crossings and won. It had come down to a simple choice between the light and the darkness. There was nothing in between.

Sean O'Malley, a coastguardsman stationed on Nantucket, was about to call it quits. He'd landed five stripers, kept one, but hadn't had a hit for over an hour, not since the tide had started falling at 8:30. Tomorrow was a workday.

234

The Wind Masters

"After I finish my cigarette," he promised.

His cast arched out past the breakers and fell through the beam of the Sankaty Head Light. It landed with a soft "plop," and Sean started his retrieve.

He didn't see the bird until it was almost abreast. He wouldn't have paid any attention except the bird was so close it almost struck his line. At first he thought it was a gull, and then he wasn't sure, and then he figured his first guess had been right.

The bird continued inland until it reached the dunes and then it collapsed. It didn't have the strength to find a safer perch, so it stayed where it was, a solitary bird, heir to a line of survivors that had sought passage and won.

In the morning it would wake to hunger and thirst. But first it would sleep. Without dreams.

THIRTY-TWO

Aquila chrysaetos
GOLDEN EAGLE

THE RAVEN SPOTTED THE DEER and the eagle on the floor of the canyon, just before the sun slipped below the stony rim of Texas's Chisos Mountains. He landed ten feet from the deer, thirty feet from the eagle. Neither protested the intrusion, and there was a reason for this. The deer had been dead for several days and the eagle, who was little more than the wasted shell of an eagle, was too close to death to notice.

The corvid cocked his head, studying the moribund raptor with calculating eyes, liking what he saw. Position prone. Eyes closed. Breathing shallow, labored. Condition emaciated. Feathers fluffed and glossless.

"Right down to the tarnished gold on your crown, eh, Princess?" he muttered under his breath.

Stalking forward, the corvid watched as the eagle strained to raise her tail . . . then muted, a vile green discharge that dribbled more than spewed from her vent, pooling atop the caked waste that lay beneath and around her. Clearly, the bird was sick.

"Feelin' a little under the weather, darlin'?" the raven invited. The question was expressed as a chortle.

At the sound of it, the eagle opened her eyes, lifted her head, staggered to her feet, and did her best to appear menacing. The raven, more amused than impressed, rasped an invective and retreated to one side in a series of hops.

The eagle managed to stand until the raven completed his retreat. Then, surrendering to gravity, she slumped to her booted tarsi and fell forward until her emaciated keel touched the earth. Only the opened wings, spread like stabilizing outriggers, prevented the bird from falling to her side.

"Old Zeus got you workin' overtime," the corvid suggested to the eagle, who had, according to legend, served as the messenger for the supreme deity of ancient Greece. The remark came out as a chuckle that was half taunt, but this time the eagle didn't respond.

"Know the problem," the raven continued, buying time with talk, studying the angles, liking the situation better all the time. "Used to work as a messenger boy for a deity myself."

The eagle didn't respond to this either. Instead she leaned forward. Opened her mouth. And her body convulsed . . . once . . . twice. . . . But the mass of food lodged in her throat would not be dislodged.

"Ahhh haaaa!" the raven rasped. "Caught you eating children again, didn't I? How many times have I got to tell you to stop stealing babies from cradles?" he taunted, dredging up one of humanity's most baseless and persistent myths concerning the Golden Eagle.

But the eagle, who had stopped convulsing, did not rise to the taunt. In fact, the bird seemed indifferent to the raven, and it was then that the raven noticed the eyes. Not only were they dull, they were lifeless.

Playing a hunch, the raven stalked to a point directly in front of the unmoving eagle.

"Odin's stronger than Zeus," he croaked.

The eagle struggled to stand once more, staring blankly forward, and the raven knew that she was blind.

"Ha! Gotcha!" the raven chortled, wondering how best to exploit the situation.

Striding back to the deer, reaching in with a daggerlike bill, the

237

"Feelin' a little under the weather, darlin'?"

The Wind Masters

raven probed the violated opening in the animal's side and skimmed away a bit of sun-ripened viscera.

"Maybe I should relieve Oedipus's sister over there of her eyes," the raven mused, letting his thoughts stray to the black side of raven humor. Eagle eyes would be a rare treat; a dainty fit for a king. Eyes that had once surveyed the world from the battle standards of conquering Roman armies. Eyes that looked with parity into the eyes of falconer kings. Eyes whose acuity is celebrated by simile and whose collective span covers almost all the land in the Northern Hemisphere — from the Canadian Arctic to the deserts of Mexico; from Scotland to Norway to North Africa and across to Russia, the Middle East, China, and Japan (the lands the Golden Eagle calls home).

"Be a bit of poetic justice plucking those eyes," the raven reasoned, building a case. "Essential element to the classical tragedy. Fall of the nobly born and all that."

"All in favor signify by saying 'EYE!'" he said to the deer, who said nothing.

But the raven was merely amusing himself. The fact was, the corvid had a healthy respect for North America's most formidable raptor, even sick and blind. He would not willingly place himself within striking distance of talons that can dispatch pronghorn antelope and have even been known to kill full-grown deer.

There was another reason for the raven's considered ambivalence. He and the eagle had a great deal in common, an association that transcended present circumstances. Both were birds who thrived in wilderness. Both were cliff-nesting species who sometimes traded nest sites. Both were masters of the air.

And both birds were clearly magical creatures — magical in the sense that they had the power to capture human imagination and use it to transform themselves in human minds. In Greek and Norse mythology, each had been the respective emissary of deities that ruled the heavens. To Native American tribes both birds were deities themselves. The tribes of the Pacific Northwest believed that it was the raven that brought life and order out of chaos. To tribes of the American Southwest, the Golden Eagle was the "Thunderbird," the creator, the embodiment of the Great Spirit himself.

GOLDEN EAGLE

The eagle and the raven enjoyed a great deal of commonality — and this made them confederates (at least in spirit). But in nature, when two creatures have much in common, their association bears a different label — competition.

Tossing his head back, bolting down the visceral morsel, the raven tried to decide what to do. He had the deer, he reasoned. He had time on his side.

"More time than Ms. Rex over there," he assessed.

And in time, well . . . who knows . . . maybe in time . . .

"Just don't turn your back on me," the raven shouted over his shoulder to his rival and soul mate.

The eagle, who was indeed blind (or nearly so), heard the taunt but was past caring. Once, the bird's awareness had extended as far as an eagle's eyes can see and as far as her seven-foot wingspan could bear her. But the death that was in her had constricted the borders of her world.

Five days ago, the planted death had robbed her of the skies. Three days ago it had reduced the world to the floor of the arroyo. Then it had taken the light out of the world, binding her to the place where for two days she had stood.

Now death was sending her spirit into retreat. As she and the raven were rivals in the sky, death and her spirit were engaged in a contest for her body — and the body was losing.

There were many things that the young eagle did not know, and would never know, about her impending death and the circumstances leading to it. She did not know, for instance, that her death was statistically sound. Among birds of prey mortality in the first year of life is high, between 50 and 80 percent, depending upon species. After two or three death-plagued years, annual mortality decreases and the odds for survival go up. Some larger raptor species, such as the Red-tailed Hawk, the Ferruginous Hawk, and the Osprey, may live 14 . . . 19 . . . even (in the case of Ospreys) 25 years! — a very long time for a bird in the wild. Plenty of time to sow the genes that are every creature's bid for immortality.

Among smaller species, the picture is not as optimistic. The annual mortality among birds the size of American Kestrels and Cooper's

The Wind Masters

Hawks remains high across all age groups. Few, if any, birds live more than 10 years — which is why many smaller raptor species become sexually active early in life and why many produce large clutches.

Had the young eagle lived to sexual maturity, a process believed to take about three and a half years, she might have lived as long as an Osprey, perhaps longer. One Golden Eagle held in captivity lived 45 years. And although she would not live — had stopped, in fact, caring about life — this is the life she might have lived had she been among the one in four Golden Eagles that survive to maturity.

In the fourth or fifth year, she would have found a mate and a territory in the Brooks Range of Alaska, the northern limit of the bird's North American breeding range. This was where she had been a nestling. This was where she would have returned.

She probably would have been recruited by an established, older male to replace a mate that had failed to return — but it is possible that she and another younger bird might have occupied a vacant territory. For Golden Eagles, this means an area ranging from 20 to 200 square miles, the boundaries defined less by the availability of prey than by the territorial limits imposed by neighboring eagles. Golden Eagles tend to be regular and regimented with regard to their zoning codes, although hunting areas at the rim of a neighboring home range are sometimes shared.

Somewhere, on the south face of a cliff — one stained red with the lichen that flourishes where raptors (and ravens) have nested before — they would have selected a ledge and fashioned or added to an existing stick nest. The ledge would offer open approach from the air, some protection from marauding predators, and strong updrafts to assist birds bearing prey. Ideally, it would be lower than prime hunting areas (so that birds burdened by prey in the extra-ground-squirrel weight class would have a downhill run).

Foxes and wolf pups fall into this category. So do the lambs of Dall sheep. But small- to medium-size mammals are the dietary mainstay of Golden Eagles. In Arctic Alaska, this means ground squirrels. Over much of the American West, this means prairie dogs, marmots, and, *particularly,* jackrabbits. Prey larger than jackrabbits is the exception, not the norm, and prey weighing more than three pounds is generally not carried more than a short distance.

The weight limit that a Golden Eagle can bear is speculative and variably dependent. But given a large eagle (one weighing eleven to twelve pounds), a favorable wind, and an elevated point of departure, it has been established that eagles can fly with prey weighing ten pounds! It is possible that they may even be able to carry more than this.

The adult eagle that would never be would have laid two eggs in late April. She would have incubated them for 44 days with help from her mate. She would have vigilantly tended her young for the first five weeks of their lives and assisted the male with hunting for five more weeks until fledging. By September, the single surviving young bird (that statistics say she would have) would have gone, just as she had gone. Also according to statistics, the bird would never return (just as she would never return).

Finding a way to die is easy if you are a bird of prey. Even among predators, life holds an assortment of death-dealing opportunities. The process can be swift or protracted. It can be natural or related to human action. It can be motivated or accidental. The young female's impending death was protracted, unnatural, and accidental. She was the victim of human unmindfulness compounded by human error.

The sad saga began on the prairies of Colorado, where several friends decided to kill a Saturday afternoon by popping the residents of a prairie dog town. They used .22 caliber rifles, equipped with 4× telescopic sights, and hollow point ammunition designed to shatter on impact.

They didn't want the carcasses, of course. They were just shooting "varmints." They also didn't trouble to police the town by throwing the carcasses down the burrows. "Let the scavengers have them," they reasoned. They actually thought they were doing wildlife a service.

The young eagle, migration taxed and hungry, had joined the host of magpies feeding on the carcasses. In the process, she ingested several small bullet fragments — most of which did not remain with the bird for long. Along with the fur and bone she ingested, they were cast up as a pellet.

The Wind Masters

But for the twenty-four hours that the fragments remained in the bird's digestive tract, they released lead, which was absorbed into the bird's system. Worse, one small fragment passed to the gizzard, where the poisoning process continued. When the bird continued her migration, she carried the incipient seed of her death inside her.

Lead poisoning is a progressive, degenerative disorder that kills birds directly or weakens them to the point that they succumb to other complications. Death can occur in a matter of days, weeks, or months.

It takes only one ingested fragment the size of the letter *o* on this page to kill an eagle. The fragment in the bird's stomach was smaller than this. Alone, it would not have brought lead levels in the bird to fatal levels. But the poisoned splinter was not so insignificant that the bird did not experience some debilitating effects.

Bathed by gastric acids, the lead leached into her system and began accumulating in the liver, the kidneys, the heart, the spleen, and the brain. Tissues began to break down. Hemoglobin levels began to fall. The bird's gall bladder became swollen, bile clogged, and her droppings turned a ghastly green.

Discolored discharge was not the only sign of the bird's illness and not the first symptom to manifest itself. Very early in the toxic process the lead began to affect the bird's nervous system. Her flight became less coordinated and more effortful. Several days after ingesting the pellet, she found herself tiring easily and perches her feet should have found easily were missed!

So was prey. Her coordination impaired, opportunities that should have resulted in kills had she been a healthy eagle were muffed. As the days passed, and as her none-too-lavish fat reserves disappeared to meet the demands of flight, she began to starve, metabolizing even the muscles that bore her aloft. By the time the poisoned bird happened upon the deer caracass, eight days after ingesting the bullet fragment, she had lost nearly 15 percent of her body weight.

She gorged herself on that deer! Tearing viscera from the animal's open side because it was easier to remove than muscle. She found it difficult to hold her balance — in fact, was forced to use her wings to brace and balance herself. But she fed until her crop was full, then flew, heavily, to a rock on the hillside — unaware that she carried her death in her.

The deer, like the prairie dogs, had been shot — but poorly. The shot had been long, the angle poor. The running animal had been hit in the haunch, and the shooter, believing that he'd missed, didn't follow up his shot to be sure.

The man was usually a better hunter than this. But it was the end of the day and he was tired. The deer had been on an opposing hillside and the terrain was bad. He saw no indication that it had been hit.

The animal had reached the arroyo before dying. The eagle had found it the next day after a coyote had worked it over. In the process of feeding, she ingested another bullet fragment.

What made the meal mortal was a combination of things. First, except for the fragments, the viscera were virtually waste free. There was no hair and few bone splinters to form a pellet, so the lead remained in the eagle's system — raising the amount of lead to lethal levels.

Second, one of the systems that lead affects is digestion. Peristaltic movement, the muscular contractions that move food through the digestive tract, is suppressed. Though the bird had fed, she could not efficiently process the food she had ingested. Instead of passing cleanly through her stomach, the half-digested matter lodged in the digestive tract (along with the lead).

Her body continued to starve. Her system continued to absorb higher levels of lead. And over the course of several days, she continued to feed — adding more and more food to the stagnant mass that was turning septic in her throat. It was the rotting food, as much as the lethal dose of lead, that would kill her.

Some might argue that the eagle's death would make little difference. Golden Eagles, after all, are a very common species, particularly in the West. In North America an estimated 70,000 adults space themselves across grasslands, sagebrush flats, deserts, tundra, taiga, and sparsely wooded areas, especially where the terrain is contoured by mountain slopes. In the East, the bird has been extirpated as a breeder, but birds heralding from New Brunswick and Quebec winter along Atlantic coastal marshes and are easily seen from known

The Wind Masters

vantage points during migration — October and November; March and April.

Some ranchers and sheepmen might even celebrate the bird's demise. For as long as shepherds have herded sheep, Golden Eagles have been vilified for killing lambs, and while the accusations are not unfounded, they are exaggerated. In most cases, eagles found feeding on lambs are exploiting stillborn animals, not feeding on kills. But this unconvincing truth (and protective federal legislation passed in 1963) has not prevented livestock owners from trying to protect their profit margin and killing eagles on sight. At times, in places, the killing of eagles has been elevated to slaughter. In the 1950s and 1960s a group of Texas ranchers declared war on *all* wintering eagles, systematically shooting them from fixed-wing aircraft. Twenty *thousand* birds were estimated to have been destroyed — almost eight times the total number of migrating Golden Eagles recorded at Hawk Mountain, Pennsylvania, during the course of its sixty years of record keeping.

Targeted persecution is not the only way that Golden Eagles suffer at the hands of our species. Golden Eagles have been caught in leg-hold traps set around carcasses as a means of controlling predators. They have also succumbed to secondary poisoning by feeding on rodents that have eaten, but not yet died from eating, poison-treated grain.

Electrocution is another major cause of Golden Eagle mortality. In treeless regions, roadside utility poles are commonly used as perches. As it lands, the bird's seven-foot wingspan is wide enough to complete an electric arc between parallel wires, with electrifying and tragic results.

Highways constitute an attractive nuisance to eagles in other ways. Road-killed carrion draws hungry young birds who know nothing about automobiles and learn about them too late, and highways are also the habitat of choice for gun-toting entertainment seekers who find raptors sitting on telephone poles more rewarding targets than highway signs.

Malicious shooting is by far the greatest known cause of death among birds of prey. It is sad. It is stupid. It is also illegal in the

United States. But the shooting of raptors is commonplace over much of the world and across the United States, too.

It is impossible to determine what percentage of the deaths of birds of prey each year are a direct or indirect consequence of human actions. Banding studies are inherently biased, overrepresenting birds whose paths intercept our own, and long-term population studies using marked individuals may offer accurate insights into mortality as it relates to that population of that species but not whole populations.

Nevertheless, on the basis of evidence accumulated through banding, from the records maintained by wildlife rehabilitation people, and from studies involving raptor corpses found in the field, it is evident that human actions are a very common cause of raptor mortality. Deaths resulting from shooting, intentional and secondary poisoning, electrocution, pesticides, and auto-related causes are probably as significant as starvation, accident, predation, and disease. For some species, humans may even be the *most* determining means to their end.

The eagle leaned forward, trying to retch again, but nothing came from her bill but a drop of clear, thin liquid. She tried again and failed again, but this time her efforts touched off a convulsion that shook her body.

The raven, who had nearly finished feeding, stepped back to watch the eagle's death throes with a dispassionate eye. Turning back to the deer, he reached deep into the cavity and extracted a fine piece of liver.

"Nevermore," the bird rasped.

The eagle was on her side now. Her breaths serial, shallow, and irregular. The bird's universe was shrinking fast. Her life imploding like a cold, spent star.

It was as though the bird and the world were overlapping planes — one fixed and firm, the other riding above — and the plane that was the bird was growing small and restless. Though it remained flush with the plane that was the earth, their point of convergence was diminishing, growing . . .

Smaller! As small as a ground squirrel from a quarter mile high. As small as the terror-filled vessel that is a rabbit's eye. As small as the first pip in the egg that was the first threshold the eagle had crossed. As small as the bitter piece of lead that would ferry her across the last.

The Wind Masters

The mote that was the essence of the bird began sliding toward the rim of the plane. Hesitating at the edge, clinging like a bird poised for flight, there was a moment when it seemed the two planes would remain together after all . . . and then they parted.

The visible sign of this was the body of the bird, which contorted suddenly. Its back arching; its head and tail straining to feel the touch of the other; its taloned feet grappling for a hold on a world that was gone. The bird's feathers lost their allegiance to each other, and as the bird's spirit passed through them, they rustled like wind passing through leaves.

By this time, the bird's body was just a body. It had ceased to be the bird who was falling; falling as mortal eagles only dream of falling. And as she fell, she stopped diminishing and started expanding once again, spreading wings that reached across an unencumbered horizon; completing an arc and becoming herself complete.

She had become both message and messenger. She was both cause and effect. She had shed her mortal form and entered the final stage of every creature's development, becoming again what she had always been.

She *was* the Great Spirit. And she was free.

The raven, who was struggling with a particularly tenacious piece of pericardium, heard the death rattle of the eagle, paused, but did not turn to watch.

"Poe beats Tennyson," he muttered to the hole in the side of the deer. Then he resumed eating, not stopping until he was full.

247

THIRTY-THREE

Gymnogyps californianus
CALIFORNIA CONDOR

THE ADULT CONDOR HALF-OPENED A WING, exposing the silver lining within and the numbered white wing tag above. More like a young mother tending her young than a bird engaged in a daily routine, she reached down with her bill and began preening her outermost flight feathers. In the process, one of the great airfoils was dislodged.

For a moment it remained suspended, impervious to gravity and time. Then it began to fall, surrendering to the fate of things that pass the span of their existence. It almost reached the earth, *would* have reached the earth but for the intercession of a heat-driven gust that raced up the canyon.

The whirlwind gathered the feather, bearing it aloft — a two-foot shaft, larger and heavier than many birds that fly. In concert with nine other primaries and twenty secondaries, it could support the 20- to 23-pound weight of the California Condor, this feather. When Lewis and Clark went exploring, the sound made by air passing through these slotted wind instruments could be heard from the Columbia River to Baja California. A century and a half later, the wind song of the condor could hardly be heard beyond the mountains of Los Padres National Forest, north of Los Angeles.

The feather continued climbing. Past the perch. Past the bird. Reaching for the pollution-thickened sky above Los Angeles. After two years of service it was the first time the feather had experienced the birthright of a California Condor, free, soaring flight, and it was making the most of it, climbing . . .

Like a dream, like a hope, past the observation window, past the wooden nest box, past the other condors closeted in the box canyon. Climbing . . . until it reached the steel net that marked the top of the condor's enclosure. Then it stopped.

The condor did not see the feather fall. She had returned to her preening, as though she were a wild, free-flying condor and such care for flight feathers truly mattered. For nine years she had lived in 40 × 40 × 100-foot enclosures, one of the birds of the California Condor Recovery Project — a joint operation involving the U.S. Fish and Wildlife Service, the U.S. Forest Service, the U.S. Bureau of Land Management, the California Department of Fish and Game, the Los Angeles Zoo, the San Diego Wild Animal Park, and independent raptor biologists.

For nine years, she had performed the everyday tasks associated with being a California Condor. Sitting, preening, bathing, sleeping, and feeding more or less regularly — as regularly as scavengers heralding from the age when Pleistocene mammals ranged across North America are supposed to feed. Even periodic privation is part of a condor's care package, so closely does the life of a captive condor replicate the life she would have known in the wild.

Like condors in the wild, birds that are part of the recovery program play, exercise their wings. Like wild condors, captive condors court, mate, and lay eggs so precious that they cannot be left to the care of parents or chance.

In fact, there is just one thing that captive condors bearing the white wing tag cannot do. That is fly free — and there is reason for this. In all the world, counting adults and young, there are fewer

249

than one hundred California Condors. Among these captive birds, whose numbers are divided between the Los Angeles Zoo, the San Diego Wild Animal Park, and the World Center for Birds of Prey in Boise, Idaho, are fourteen birds known as "Founders," the birds that are the last wild California Condors on the planet.

Housed in these Founders is the genetic dowry of the entire species; all that remains; all that can be passed on. At one time it was hoped that at least some of these Founders might remain in the wild — an unbroken link to the past; a pool of experience that birds released into the wild could draw from — but it was not to be. After decades of attrition, after nine of the fifteen condors not yet held in captivity were lost between November 1984 and April 1985, it was clear that this optimistic strategy was bankrupt. If the condor was to have any future, it was imperative to bring all condors into protective custody and the genetic pool.

The female, known as "UN1," was one of these.

She finished her preening, straightened, and turned toward the sun, opening both wings, exposing them to the warming rays. It was winter and, for Los Angeles, chilly — though not as chilly as the cliffs of the Sespe Condor Sanctuary, the principal nesting area for the birds during the last half of the twentieth century. She held them this way for several minutes, the wings wilting slowly, then turning, she spread them once again, letting the sun warm the upper surface, too, exposing the white wing tag emblazoned with the number 13. Between the preening, stretching, and sunning, the morning ritual could consume five hours. Even in the wild, without a barrier between them and the sky, California Condors spend most of their day perched.

With or without a white wing tag, it was clear that the bird was an adult California Condor. The heads of young condors are bare and gray. Hers was bare and pink, with eyes that glowed like peat trapped in amber and a bill that was covered in wrinkled flesh almost to its hooked and horn-colored tip. Wrapped in a loose, black cloak whose trim was a feather ruff, she resembled an aging dowager — but the fact was that her age and much else about her were a mystery. UN stands for "unknown." One day, during the period when the last

The Wind Masters

free-flying condors were being intensely studied, the female had just turned up; her origin unknown.

UN1's "mate," perched at the far end of the enclosure, was better known to condor researchers and his history was illustrious. His name was Sequoia. He was the last California Condor chick to be hatched in the wild and his name bears tribute. The nest site was a burn cavity high up along the trunk of a giant sequoia. Caves are the preferred nest sites of condors, who deposit a single egg a season (and sometimes every other season). This rate of productivity is low by the standards of most birds but adequate to offset the normal rate of mortality in a long-lived species such as the California Condor — so long as other factors do not enter the equation and tip the balance.

Completing her morning ritual, UN1 left her perch and flew toward Sequoia, who responded by hiding in the wooden nest box — a simulated cave. The distance was 82 feet. She reached it in six flaps. Turning, she flew back to the original perch, then to the ground.

It was midwinter, breeding time for California Condors, and she was restless. It was the only time that she really missed the skies — missed particularly the pair flights, in which courting condors twist and turn like empathetically attuned figure skaters. At other times, flying was a necessity — analogous to humans commuting to work.

Like most scavengers, California Condors fly to forage or to relocate to some place where the foraging will be more successful. In the enclosure her food needs were well attended to, her need to search for the carrion that supported her and her kind preempted. So she almost never thought of flight now. Except during the breeding season. *And at night. When the dreams came.*

It almost always happened on a full moon, these dream flights. When the light made the other animals in the zoo a little crazy. When the strength of moonbeams cast shadows as real as substance.

Maybe moonlight exerts some power over species fading into the twilight of their existence. Or maybe the sound of the zoo's large land mammals — lions, elephants, camels — unlocked doors in her ancestral memory that carried her back 11,000 years, when the ancestors of these creatures roamed North America and condors ruled the skies from the Atlantic to the Pacific.

CALIFORNIA CONDOR

On these nights, in her dreams, she felt her wings unfold and gravity disappear. On these nights, in her mind, the moon's silver light turned to gold and she would set a course for the sun as she had in her youth. As manifest as the past, she heard again the wind song of her flight and regarded the planet from 6,000 . . . 7,000 feet . . . traveling 100 miles in one Earth hour. . . .

Just as her progeny would soar in the future. Just as young condors were soaring, *again,* in the hills above Los Angeles — the first birds successfully released by the Recovery Project. Beyond her cage, in a utilitarian array of buildings and trailers, was the future of her species. In this facility, and the sister facility in San Diego, the world's population of California Condors was being restored to self-perpetuating levels. Some of these incipient Wind Masters were in incubation chambers, where they would remain for 57 days — the time it takes condors to develop and hatch. Others, already hatched, were in carefully monitored chambers, where for two months they would be tended by technicians steeped in the skills of condor virtual reality.

The feeding of chicks is manipulated by Condor-like hand puppets. The natural sounds of the California coast are piped into their confines — surf and birds and barking seals by day; crickets and frogs by night. Human contact is minimized to prevent young birds from regarding our species as a source of benefit. Current circumstances notwithstanding, history has shown just the opposite to be true.

From brooding chambers the birds are transferred to eight-foot-square play pens and then, after six months, after fledging, to flight cages, where they associate with other condors and learn the social skills of their kind. Most of these birds are destined for release.

The techniques required to save a species from extinction did not come ready-made. Most had to be developed from scratch and at great cost. The first eggs successfully incubated were taken from the wild when it was discovered that, once relieved of their first egg, condors will lay a second. This desperate piracy more than doubled the number of eggs wild condors were producing naturally and, along with several chicks brought into protective custody, helped to bring the flock of captive condors to twenty-seven birds when the last wild condor was captured in 1987.

The Wind Masters

After the onset of the captive breeding program, the Founders began producing eggs within the monitored confines of the Los Angeles Zoo and San Diego Wild Animal Park. Four chicks hatched in 1989; five in 1990; twelve in 1991 — a total of sixty-two by 1995. The number of condors left on earth is now eighty-nine birds — a threshold that has not been crossed in this half of this century. Nineteen young condors have been released — the shock troops of a carefully calculated campaign whose ambition is to restore the birds to their former stronghold and to establish populations in parts of Arizona and New Mexico to better the odds of survival.

The Condor Recovery Project has encountered setbacks, of course. The world is a more treacherous place for young condors than it was 11,000 years ago, and without the guidance of older birds, newly released condors have no other recourse but to go out and get their beaks bloody. Some of the birds released into the wild have been lost — four to collisions with power lines; one to drinking from a pool of antifreeze lying in a parking lot.

But the Condor Recovery Project has faced setbacks before and greater obstacles than the challenges inherent in hacking inexperienced condors. These obstacles include opponents who have argued against the undertaking almost from the onset.

Some opposed the program on the basis of cost. Some because protection efforts might potentially conflict with the rights of landowners or interest groups who feared restrictions on their use of public lands. Others saw a hands-on effort to save the condor as an unwarranted risk that might accelerate the demise of the bird. A surprising number believed that restoration efforts constituted an intrusion into nature.

Extinction, the laissez-faire advocates ventured, was a natural process. The condor, they asserted, was a species whose time had come and gone. Efforts to preserve it were, therefore, contrary to nature's dynamic principles.

This argument has seductive appeal and not a few environmental organizations embraced it. It evokes the support of "science" and conjures the notion of "inviolate wilderness" — both of which are counted among our culture's most powerful totems. But the hands-

253

off argument was faulted on the facts. The truth is that there is no evidence that the California Condor population was declining until the Manifest Destiny of North America's human population conflicted with the destiny of the bird. At the end of the Pleistocene Epoch, when large land mammals disappeared over much of North America, the California Condor did not follow suit. The condor staged a strategic withdrawal to the coastal mountains of the Pacific, where it survived nicely — navigating the updrafts off the coastal mountains; foraging on the remains of marine mammals such as whales, seals, and sea lions, and scavenging large ungulates (deer, elk, mountain sheep).

Food was plentiful. The habitat secluded and supportive. The birds settled in to sit out the millennia.

Then European settlers arrived. The balance that the California Condor had struck with the Universe went south.

There is no need to offer a detailed account of the condor's two-century retreat from homeostasis to near extinction. The decline *has* been studied. The causal agents *have* been documented. The blaming finger points squarely at our species. But in quick summary, shooting would seem to have been the primary cause of the bird's demise, though too-late-on-the-scene evidence suggests that lead poisoning may have exacted as high a price upon the birds as lead impact. Three of the last condors to die in the wild were the victims of lead poisoning. The source of the metal was bullet fragments ingested while birds fed upon the remains of gunshot animals.

In the latter part of the nineteenth century, measures to eliminate grizzly bears in California by using poisoned carcasses probably eliminated condors, too. And the impact of science, so often called upon to restore the unrighted balance but too often overlooked as a source of blame, should not be overlooked here. No less than 288 birds and 71 eggs were collected between 1792 and 1976 — more birds than are known to have been killed by shooting (although scientists, whose discipline it is to amass evidence, are far more likely to keep records than those who kill condors for less defensible reasons).

Accidental poisoning . . . accidental drowning (in a cattle trough!) . . . power line collisions . . . capture for display (as pets and as zoo attractions) — all levied a price against the dwindling number of

The Wind Masters

birds. DDT, which was widely applied to the agricultural lands within the condor's realm, also affected the condors, resulting in a 30 percent reduction in condor eggshell thickness and documented nest failures.

But the bird's problem was not productivity. In fact, even into the last days of efforts to maintain a wild condor population, reproduction rates remained at levels considered adequate for a large, carrion-eating vulture species. The problem was mortality among adults. As a long-lived species, condors could live with low levels of recruitment. What they could not absorb was the accelerated rate of attrition to the ranks of adult birds that our species, wittingly and unwittingly, imposed upon them.

Those whose opposition to the Condor Recovery Project rests on the grounds that humans should not interfere with the natural process of a species are overlooking one very salient point. We *are* involved and we *have been* involved with the California Condor for two hundred years.

What we have not been, until now, is *responsible*.

The question is not whether condors can survive in this world. They can. They've proved their staying power. The question is whether condors can survive in this world dominated by humans. We have, to the planet's great loss, always taken upon ourselves the privilege of asserting ourselves at the expense of other living things. The imminent demise of North America's greatest soaring bird is a test of our wisdom and our resolve. It forces the question: Are we finally mature enough to accept responsibility for our actions and to try to accommodate the needs of other living things?

The effort of the California Condor Restoration Team says that we are willing to be that responsible. It bodes well for California Condors, and for all species that strive to coexist on the planet. Including our own.

UN1 spent the afternoon as she spent most afternoons, resting. As the shadows lengthened, as the noises of the world beyond her world diminished, she flew once more to the perch near the nest box. This is where she would spend the night. All around her the other condors were going through similar preparations.

Sleep did not come quickly to the bird. Her energy expenditures

255

"*She was soaring. As she had in her youth. As her progeny would soar in the future.*"

256

The Wind Masters

had been minimal and so was her need for sleep. For some minutes she remained awake. Only gradually did her eyes close . . . and remain closed.

She slept standing up, as hawks and vultures do, with her head hunched, not tucked. She woke easily and often, stirred to wakefulness by disparate sounds — a scurrying mouse, a condor shifting on its perch, the sudden sound of a radio as the door to some late-working zoo employee's enclave was opened and closed.

But the other sounds, the sounds of the animals in their confines and the busy hum of the world beyond, did not disturb her. She had grown used to them. They were now as familiar as the sound of crickets and surf.

She did not hear the clamor grow as the moon climbed over the hillside. She did not specifically hear how the lions roared their challenge, and the baboons their retort — nor how the moon songs of creatures whose ancestors had roamed this land when condors ruled the skies joined in chorus, paying homage to their heritage with all the tribute heritage asks, survival.

She was hearing by that time the wind song of a condor's wings, and seeing again the opposing horizons through slotted wing tips. She was soaring. As she had in her youth. As her progeny would soar in the future, bearing her genes aloft and their heritage forward.

Bibliography

American Ornithologists' Union. *Check-list of North American Birds.* 6th ed. Lawrence, Kans.: Allen Press, 1983.

Angell, Tony. *Ravens, Crows, Magpies and Jays.* Seattle and London: University of Washington Press, 1978.

Arnold, Caroline. *On the Brink of Extinction — The California Condor.* San Diego: Harcourt Brace Jovanovich, 1993.

Bent, Arthur Cleveland. *Life Histories of North American Birds of Prey.* 2 vols. National Museum Bulletins nos. 167 and 170. Washington, D.C.: Smithsonian Institution, 1937 and 1938.

Brown, Leslie, and Dean Amadon. *Eagles, Hawks and Falcons of the World.* 2 vols. Feltham, England: Hamlyn House, 1968.

Cade, Tom J. *The Falcons of the World.* Ithaca, N.Y.: Comstock/ Cornell University Press, 1982.

Choat, Ernest A. *The Dictionary of American Bird Names.* Boston: Gambit, 1973.

Clark, William S., and Brian Wheeler. *A Field Guide to Hawks North America.* Boston: Houghton Mifflin, 1987.

Dunne, Pete, David Sibley, and Clay Sutton. *Hawks in Flight.* Boston: Houghton Mifflin, 1988.

Friend, Milton. *Field Guide to Wildlife Diseases.* Vol. 1. *General Field Procedures and Diseases of Migratory Birds.* Washington, D.C.: U.S. Department of the Interior Fish and Wildlife Service, Resource Publication 167, 1987.

Bibliography

Gill, Frank B. *Ornithology.* New York: W. H. Freeman and Company, 1990.

Greij, Eldon. "Condors Try Out New Digs." *Birder's World* 8, no. 5 (1995): 9.

Harrison, Colin. *A Field Guide to the Nests, Eggs and Nestlings of North American Birds.* Glasgow, Scotland: William Collins Sons & Co., Ltd., 1978.

Johnsgard, Paul A. *Hawks, Eagles and Falcons of North America.* Washington, D.C.: Smithsonian Institution Press, 1990.

Kerlinger, Paul. *Flight Strategies of Migrating Hawks.* Chicago: University of Chicago Press, 1989.

Kiff, Lloyd. "To the Brink and Back." *Terra* 28 (1990): 7–18.

Langley, Lynne. "Swallowtails of the Francis Marion." *South Carolina Wildlife:* 7–10.

Newton, Ian. *Population Ecology of Raptors.* Vermillion, S.Dak.: Buteo Books, 1979.

Palmer, Ralph S. *Handbook of North American Birds.* 2 vols. New Haven, Conn.: Yale University Press, 1988.

Pattee, Oliver H., Stanley N. Wiemeyer, Bernie M. Mulhern, Louis Sileo, and James W. Carpenter. "Experimental Lead-Shot Poisoning in Bald Eagles." *Journal of Wildlife Management* 45, no. 3 (1981): 806–810.

Poole, Alan F. *Ospreys: A Natural and Unnatural History.* Cambridge: Cambridge University Press, 1989.

Rand McNally Cosmopolitan World Atlas. Chicago: Rand McNally & Company, 1985.

Ratcliffe, Derek. *The Peregrine Falcon.* Vermillion, S.Dak.: Buteo Books, 1980.

Robbins, Chandler S., Bertel Bruun, and Herbert S. Zim. *Birds of North America.* 2d ed. New York: Golden Press, 1983.

Salzman, Erik. "Sibley's Classification of Birds." *Birding* 25, no. 6 (1993): 446–455.

Skutch, Alexander F. *Birds Asleep.* Austin, Tex.: University of Texas Press, 1989.

Snyder, Noel, and Helen Snyder. *Birds of Prey — Natural History and Conservation of North American Raptors.* Stillwater, Minn.: Voyageur Press, 1991.

Strahler, Arthur N. *Physical Geography.* New York: John Wiley and Sons, 1975.

Sykes, Paul W., Jr. "A Closer Look: Snail Kite." *Birding* 26, no. 2 (1984): 118–122.

Terres, John K. *The Audubon Society Encyclopedia of North American Birds.* New York: Alfred A. Knopf, 1956.

Watson, Donald. *The Hen Harrier.* Hertfordshire, England: T. & A. D. Poyser Ltd., 1977.

Wetmore, Alexander. *Water, Prey and Game Birds of North America.* Washington, D.C.: National Geographic Society, 1965.

PETE DUNNE is the author of several acclaimed books, including *Pete Dunne on Bird Watching, Hawks in Flight, The Feather Quest,* and *Tales of a Low-Rent Birder.* He has written columns and articles for many birding magazines and for the *New York Times.* The vice president of the New Jersey Audubon Society and the director of its Cape May Bird Observatory, he has received the American Birding Association's Roger Tory Peterson Award for Promoting the Cause of Birding.

PETE DUNNE ON BIRD WATCHING
The How-to, Where-to, and When-to of Birding

> "Dunne is like an older friend letting you in on his experience and some of his hard-won secrets." — Eric Salzman, *Winging It*

In this lively and authoritative guide, Dunne identifies the skills and tools available to people with any amount of interest, great or small, in bird watching. He outlines the basic, simple steps that can make a birder out of the most casual observer. **HOUGHTON MIFFLIN PAPERBACK ISBN 0-395-90686-5**

THE FEATHER QUEST
A North American Birder's Year
Foreword by Roger Tory Peterson

> "Makes you want to grab your binoculars and head outdoors."
> — *Philadelphia Inquirer*

Pete Dunne and his wife, Linda, spent an extraordinary twelve months traveling North America in search of birds. Dunne's wisdom and wit make this colorful chronicle of their odyssey both an irresistible travel book and an information-packed introduction to the joys of birding.

A MARINER BOOK ISBN 0-395-92790-0

HAWKS IN FLIGHT
The Flight Identification of North American Migrant Raptors
(with David Allen Sibley and Clay Sutton)

> "A landmark." — Roger Tory Peterson

This unique identification guide shows how to recognize twenty-three of the most common diurnal raptors of North America the way we recognize friends at a distance: by body shape, by the way they move, and by the places they are most likely to be seen. **HOUGHTON MIFFLIN PAPERBACK ISBN 0-395-51022-8**